Engage!
Transforming Healthcare Through Digital Patient Engagement

Editor

Jan Oldenburg, FHIMSS

Contributing Editors

Dave Chase
Kate T. Christensen, MD
Brad Tritle, CIPP

HIMSS Mission

To lead healthcare transformation through effective use of health information technology.

Printed in the U.S.A. 5 4 3 2 1

Requests for permission to make copies of any part of this work should be sent to:
Permissions Editor
HIMSS Media
33 West Monroe Street, Suite 1700
Chicago, IL 60603-5616
Nancy.vitucci@himssmedia.com

ISBN: 978-1-938904-38-7

For more information about HIMSS, please visit www.himss.org.

About the Editor

Jan Oldenburg, FHIMSS, is the Vice President of Patient and Physician Engagement in Aetna's Accountable Care Solutions organization. In this role she is responsible for working with provider partners to design and implement patient engagement strategies as well as the governance, marketing, and monitoring programs that will ensure that the provider organizations reach patient engagement goals.

Ms. Oldenburg's background includes nearly 20 years working at the intersection of consumers, healthcare and web and mobile tools. Prior to joining Aetna, Ms. Oldenburg spent seven years in Kaiser Permanente's Internet Services Group. During that time, her responsibilities included deploying the patient-facing medical record in Northern California, developing the personal health record (PHR) product plan, and, at various times, leading teams of product managers responsible for interoperability, content, languages, cost transparency, revenue cycle, and health plan features. She also led Kaiser's Meaningful Use Stage 1 workstream on empowering patients and their families.

Prior to joining Kaiser, Ms. Oldenburg was the principal in several consulting companies, specializing in strategies for using the Internet effectively in healthcare. Clients included United HealthCare, Medica Health Plan, Kaiser Permanente, Blue Cross-Blue Shield of Minnesota, and Medtronic. In the mid-1990s she was the Senior Director for Electronic Commerce at HealthPartners, a Minnesota HMO, where she developed a secure transactional Internet portal for contracted providers.

Ms. Oldenburg is currently President of the Northern California HIMSS Chapter and has been active on its board since 2003. She is an advisor to the HIMSS eConnecting with Consumers Committee. Previously she has been an active member of the HL7 PHR Workgroup as well as the CCHIT PHR Standards Workgroup. Ms. Oldenburg received her BA degree in English and Philosophy from Luther College in Decorah, Iowa, summa cum laude, and completed coursework (though not the thesis) for a PhD in English.

About the Contributing Editors

Dave Chase is the co-founder & CEO of Avado, the first complete cloud-based Patient Relationship Management system that includes a multi-provider, EHR-agnostic patient portal and Collaborative Health Record. Mr. Chase is also a regular contributor to publications such as Reuters, Forbes, TechCrunch, The Health Care Blog and The Doctor Weighs In. Earlier in his career, Mr. Chase implemented or reviewed over 100 health IT systems while working for Accenture's healthcare practice on behalf of over 20 healthcare systems. He went on to found Microsoft's health platform business which became the preeminent underlying technology for the majority of the health IT vendors and is Microsoft's most successful vertical market. Since leaving Microsoft in 2003, he has been a leader/co-founder of several successful technology startups including LiveRez, Market Leader and WhatCounts.

Kate T. Christensen, MD, is the Medical Director for the Internet Services Group of Kaiser Permanente, and is responsible for clinical oversight and strategy for the healthcare organization's patient portal, kp.org. She also serves as the Medical Director for the Kaiser Permanente Martinez Hospice Program, as well as the National Physician Lead for Patient-Centered Care for the Care Management Institute. In 2012 she was appointed the chair of the HIMSS eConnecting with Consumers Committee. A general internist as well as hospice and palliative medicine specialist, with a background in medical ethics, Dr. Christensen has published and presented in the U.S. and internationally on the topics of medical ethics, managed care, end-of-life care, and online patient engagement.

Brad Tritle, CIPP, is the Director of Business Development for Vitaphone USA, and co-founder of eHealth Nexus and Health-e Republic. At Vitaphone, he is responsible for all aspects of the Chronic Disease Management Market for remote patient monitoring services and devices. He is or has been a consultant and advisor to Informative Graphics/IGC, eHealthTrust, Cox Business, 360Vantage, ONC's State HIE Consumer Innovation Challenge, and other organizations. For three years, Mr. Tritle headed Arizona's health IT non-profit, Arizona Health-eConnection, focusing on HIE, EHR, e-prescribing, and PHR adoption. Mr. Tritle serves on the HIMSS eConnecting with Consumers Committee and the Arizona Telemedicine Council, chairs the HIMSS Social Media Task Force and Health Record Banking Alliance Business Model Committee, and hosts a weekly Internet radio show at Engage4Health.com.

About the Contributors

Natasha Burgert, MD, FAAP, is a general pediatrician in Kansas City, MO. Her thriving suburban private practice combines her love of patient care and her passion for technology. She is the social media community manager for her practice group, as well as author of a child health and parenting blog entitled KCKidsDoc.com. Her unique ability to use technology to improve patient-provider communication has been featured in many media outlets including *The New York Times,* the Associated Press, and various national and local news programs.

Nancy Burghart-Hall, CHCIO, has 20 years of health IT leadership experience, working in government, not-for-profit, and public healthcare developing strategies to align and implement technology. She is currently the CIO at MedAmerica/CEP America, a physician medical group staffing emergency, hospitalist and urgent care providers at over 120 hospital locations across the country. Ms. Burghart-Hall has taught and presented on implementing healthcare information systems, governance and strategic planning, and leveraging IT to change healthcare. She is a Certified Healthcare CIO and a member of the Board of Directors of the Northern California HIMSS Chapter. Ms. Burghart-Hall is active in patient advocacy and healthcare policy at the state and national level and her personal mission is to educate patients and caregivers in the importance of managing their personal health information.

Bianca K. Chung, MPH, is the Director of Strategy for InTouch Health Technology, a manufacturer of Class II, FDA-cleared telemedicine remote presence medical devices for active patient monitoring in high-acuity clinical environments. Ms. Chung is responsible for contributing to corporate and product strategies that lead to growth and market leadership. She previously worked at Deloitte Consulting LLC, working in provider reimbursement, EMR implementation, HIE start-up business models, and provider operations improvement. Her operations background from Emory Healthcare enabled her to have direct roles in hospital and clinic operations. She has authored and presented on cutting edge solutions for healthcare delivery, reimbursement, and healthcare technology. Her focus is on mobile health, fitness and wellness, and patient-centered care.

Karen Colorafi, RN, MBA, BScN, CPEHR, CPHIT, works as an Arizona-based consultant on various electronic health record (EHR) vendor selection, implementation, and optimization projects. She has a passion for seeing ambulatory care practices enjoy the EHR experience. She holds a fundamental belief that EHRs can increase provider satisfaction, improve practice efficiency, and enhance patient care. Her experience as both a clinical nurse and healthcare technology analyst give her unique insight into the needs of medical practices as they adopt and implement EHRs. Some of her past positions include Arizona State University's Department of Biomedical Informatics, supporting the efforts of the Arizona Regional Extension Center; Clinical EHR Nurse Specialist with the CMS Doctor's Office Quality Information Technology initiative (DOQ-IT); Director of Nursing for a large, statewide cardiology practice; and nurse in cardiac rehab, cardiology, family practice and urgent care.

David Fetherstonhaugh is a behavioral economist at IDEO. He leads IDEO teams by bringing insights from behavioral science into a human-centered approach to designing for behavior in the areas of health, nutrition, energy and financial wellbeing. Because decisions often determine behavior, Dr. Fetherstonhaugh sees architecting individual choice and shaping groups norms as integral for starting, spreading and sustaining new behavior patterns. Before joining IDEO, he led large-scale research projects within government and academia in the areas on terrorism, environmental protection and health risks. He has also founded two venture-backed companies. Dr. Fetherstonhaugh has published numerous widely cited papers and book chapters on the value of human life, persuasion, and retirement decision making. He earned a Master's in statistics from Stanford University and a PhD from its department of psychology.

Susannah Fox is an Associate Director of the Pew Research Center's Internet & American Life Project, where she studies the cultural shifts taking place at the intersection of technology and healthcare. Her research has documented the social life of health information, the concept of peer-to-peer healthcare, and the role of the Internet among people living with chronic disease. Ms. Fox contributes to two healthcare blogs, e-patients.net and susannahfox.com, and she is an active member of the health geek tribe on Twitter. Ms. Fox is the former editor of the website for *U.S. News & World Report* magazine and was a researcher during the start-up phase for RealNetworks. Ms. Fox graduated from Wesleyan University with a degree in anthropology.

Adam H. Greene, JD, MPH, is a partner in the Washington, DC, office of Davis Wright Tremaine and co-chair of its Health Information Group. Mr. Greene primarily counsels healthcare providers, technology companies, and financial institutions on compliance with the HIPAA privacy, security, and breach notification rules. Previously, Mr. Greene was a regulator at the U.S. Department of Health and Human Services, where he played a fundamental role in administering and enforcing the HIPAA rules. At HHS, Mr. Greene was responsible for determining how HIPAA rules apply to new and emerging health information technologies and was instrumental in the development of the current HIPAA enforcement process. Mr. Greene is the Chair of the HIMSS Cloud Security Workgroup and is a frequent speaker and author on health information privacy and security issues.

Glen Griffiths, MSc, MBCS, an eHealth technologist/strategist, has worked in the healthcare arena since 2002 with a clear focus on delivering interactive, public, and patient online health information solutions. Since 2007 he has worked closely with Dr. Amir Hannan, coproducing the practice-based web portal, www.htmc.co.uk, and other initiatives, promoting partnership working between clinicians, patients and carers. Mr. Griffiths' previous company, interactivHealth plc, where he held the position of Chairman & CEO, delivered the purpose-built CMS upon which www.htmc.co.uk resides. Mr. Griffiths is the immediate past Vice-Chair of the British Computer Society Primary Health Care Specialist Group and holds an MSc in Information Technology from University of Liverpool (2007). Originally trained in E&EE, he also undertook post-graduate studies in Strategic Management from the University of Reading. (Twitter: @griffglen.)

Amir S. Hannan, MBChB, MRCGP, is a General Practitioner (GP) at Haughton Thornley Medical Centres in Hyde, UK. Developing a "Partnership of Trust" between patient and clinician, the practice has enabled over 1,900 citizens (17% of practice population) to access their full GP electronic health record online, helping them to self care and gain a better understanding of their health via www.htmc.co.uk, which Glen Griffiths also helped to co-produce. Presently Dr. Hannan is the Primary Care IT lead and Map of Medicine clinical lead for NHS North-West, and a member of the Health Informatics Clinical Advisory Team and Clinical Director for Patient Access to GP Online Services Advisory Group, NHS North. (Twitter: @amirhannan.)

Peter Hudson, MD, is a physician and entrepreneur with over 15 years' experience founding and growing healthcare-related businesses. His focus has been on creating efficiencies within the healthcare delivery system, and empowering healthcare consumers with technology. Dr. Hudson has been a serial entrepreneur with four exits (three to public companies), a healthcare investment banker with lots of sell-side and buy-side experience, and has served as the managing partner of large emergency medical practices. He was listed as one of the 12 Entrepreneurs Reinventing Healthcare in *CNN Money* in 2012. Dr. Hudson is a frequent speaker on healthcare and technology. iTriage was founded in 2008 and acquired by Aetna in 2011, and represented the first significant exit in the mobile health space. Dr. Hudson has practiced emergency medicine in a wide range of environments, from country trauma centers to non-profit community hospitals. In addition, he has practiced around the world in many countries, including Nepal, Guatemala, and Kenya.

Deven McGraw, JD, MPH, LLM, is the Director of the Health Privacy Project at CDT, where she focuses on developing and promoting policies that ensure individual privacy as personal health information is shared electronically. Ms. McGraw is active in efforts to advance trusted implementation of health information technology. She serves on the Health Information Technology (HIT) Policy Committee, a federal advisory committee established in the American Recovery and Reinvestment Act of 2009, and chairs its Privacy and Security Tiger Team. She also serves on the Leadership Council of the eHealth Initiative and is on the Steering Group of the Markle Foundation's Connecting for Health multi-stakeholder initiative.

David W. Moen, MD, is President of Bluestone Solutions, which is a healthcare strategy, care model development, and physician leadership consultancy based in Stillwater, MN. Dr. Moen worked within the Fairview Health System in Minneapolis for 25 years as an emergency physician and physician leader. Over the last four years he led development of the Fairview Health Network (FHN) as President and CEO of Fairview Physician Associates and led care model innovation as Executive Medical Director of Innovation. In 2012 Fairview was awarded four innovation awards for work of his teams achieving triple aim results. Dr. Moen also helped develop the Fairview Physician Leadership Academy and served on its faculty for five years. Dr. Moen has been a featured speaker in several venues nationally, and has written for and his work has been written about in several local and national publications. Leadership roles include board positions within the Fairview Health System, physician organizations including Minnesota Chapter of the American College of Emergency Physicians, and community organizations including the Center for Spirituality and Healing at the University of Minnesota.

Eve Phillips, M.Eng, MBA, is the CEO & Co-Founder of Empower Interactive, Inc. Ms. Phillips is an experienced Silicon Valley product builder, entrepreneur and investor with expertise in assembling diverse teams to solve hard problems. She started her career as a product manager at Trilogy, working on new ventures. From there, she held product, strategy and business development roles in technology businesses including Microsoft, Zazzle and eCert. Ms. Phillips has also been a member of the investment teams at Amadeus Capital Partners, Vector Capital, and Greylock Partners. Prior to co-founding Empower, she was a Research Affiliate in the Synthetic Neurobiology Group at the MIT Media Lab. Ms. Phillips holds an MBA from Stanford University and an SB and an M.Eng in Computer Science from the Massachusetts Institute of Technology.

Jane Sarasohn-Kahn, MA, MHSA, is a health economist and management consultant who works with healthcare stakeholders at the intersection of health and technology. She founded THINK-Health, a strategic health consultancy, after spending a decade as a healthcare consultant in firms in the U.S. and Europe. Ms. Sarasohn-Kahn's client base spans the broad range of stakeholders in health, including technology, bio/life sciences, providers, plans, financial services, consumer goods, advertising and communications, public sector and not-for-profit organizations. She specializes in environmental analysis, scenario and strategic planning, forecasting, and health policy analysis. Ms. Sarasohn-Kahn also writes the Health Populi blog. She sits on several advisory boards and is a frequent collaborator with the California HealthCare Foundation. She holds an MA (Economics) and MHSA (Health Policy) from the University of Michigan.

Aaron E. Sklar leads design strategy for Healthagen's portfolio of health-tech start-ups. With over 15 years' experience in design leadership, he had guided over 75 organizations along their journey of innovation. His background in human factors permeates a foundation of empathy for people at all aspects of the process. Empathy for end users, team members, clients and key stakeholders are all critical for recognizing opportunities and ensuring successful implementation. Health and wellness initiatives have always been at the forefront of Mr. Sklar's professional goals—leveraging capabilities in innovation and empathy which are uniquely suited to tackling the behavior change challenges that underlie innovation in the health arena.

Jonathan Wald, MD, MPH, FACMI, is the Director of Patient-Centered Technologies in the Center for the Advancement of Health Information Technology at RTI. A research scientist experienced in clinical care, systems design, process redesign, and product management, his publications and work in academic and corporate settings have focused on patient computing, provider health IT and workflow, confidentiality, technology adoption, and strategy. For over ten years he led the patient portal team at Partners HealthCare in Boston, creating Patient Gateway, securely connecting patients with thousands of physicians over the Internet. Dr. Wald was elected as Fellow, American College of Medical Informatics, and serves on the Harvard Medical School faculty. He received an MD from Brown University, an MPH from Harvard, and a BA from Dartmouth College.

Deborah C. Wells, MS, CPHIMS, is an Information Systems Strategic Consultant at the Children's Hospital of Philadelphia (CHOP). At CHOP, Ms. Wells is responsible for developing clinical and business systems strategy and for technology planning. In addition, Ms. Wells provides internal consulting services on a wide range of healthcare IT matters. Ms. Wells holds a Master of Science in Computer Information Systems degree from Boston University and an undergraduate degree from Tulane University. Previous to working at CHOP, Ms. Wells was Director of Classroom Technology at the Wharton School of the University of Pennsylvania.

Susan Woods, MD, MPH, is an internist and consumer informaticist at the Portland VA Medical Center and Associate Professor at Oregon Health & Science University, Departments of Medicine and Medical Informatics. At the Veterans Health Administration, Office of Informatics & Analytics, she brings the Patient Voice into the design of health IT services, focusing on shared health records and patient-generated data. Dr. Woods is a principal investigator in the eHealth QUERI, a VA virtual research consortium. Previously on the boards of the Society for Participatory Medicine and the Society of Behavioral Medicine, she promotes consumer health technology as a way to engage patients and families in their care, and enhance the patient experience. She encourages innovation and participatory care on her blog, Shared Health Data.

About the Case Study Interviewees

Kenneth Adler, MD, MMM, is a practicing family physician, and President and Medical Director for Information Technology of Arizona Community Physicians (ACP), a 126-physician, primary-care medical group in Tucson, Arizona. His three-physician office has twice been recognized by NCQA as a Level 3 Patient-Centered Medical Home. His practice has employed an EHR since 2004 and a patient portal since 2008. He holds an MD from the University of Pennsylvania, a Master of Medical Management degree from Tulane University, and a certificate in Healthcare Information Technology from the University of Connecticut, and is a Certified Professional in Health Information and Management Systems. Additionally, he is a HIMSS Fellow. For a number of years Dr. Adler has spoken nationally and internationally on health information technology and he has authored over 20 articles and book chapters on EHRs and related health information technology in peer-reviewed journals and HIT texts.

Jen Brull, MD, FAAFP, is a family medicine physician in solo practice in Plainville, Kansas. She obtained her medical degree from the University of Kansas School of Medicine and completed family medicine residency training in Topeka, Kansas. This is her eleventh year of private practice in a rural area. Dr. Brull was the first physician to attest to Meaningful Use in Kansas and is a physician leader in electronic record adoption, change management and health information exchange. Her special interests include quality improvement, social media, and preventive medicine.

Joseph A. Cafazzo, PhD, PEng, is Lead for the Centre for Global eHealth Innovation, a state-of-the-art research facility devoted to the evaluation and design of healthcare technology, hosting 70 researchers and staff. Dr. Cafazzo has been an active researcher of the use of technology to facilitate patient self-care of complex chronic conditions. He has advised and conducted research for public sector policy makers and private sector medical technology companies on the design and safety of technology in healthcare. He is Assistant Professor, Faculty of Medicine, University of Toronto, where he teaches and conducts research in the areas of human factors, clinical engineering, and health informatics. In 2010 he was the recipient of the Career Scientist award by the Ontario Ministry of Health and Long Term Care.

Dave deBronkart, known online as e-Patient Dave, beat stage IV cancer in 2007 and became a blogger, keynote speaker and health policy advisor. He is today a leading spokesman for patient engagement, attending over 150 conferences and policy meetings internationally in the past two years. He serves as volunteer co-chair of the Society for Participatory Medicine. e-Patient Dave has appeared in *Time*, *U.S. News*, *Wired*, *MIT Technology Review*, and the *HealthLeaders* cover story, "Patient of the Future." In 2009 *HealthLeaders* named him and his doctor to their annual list of "20 People Who Make Healthcare Better," and in 2011 his TEDx talk (http://on.TED.com/Dave) went viral globally: volunteers have added subtitles in 25 languages. Its tagline is his appeal: "Let Patients Help."

Lucien Engelen is Head of the Regional Emergency Healthcare Network, Advisory to the Executive Board and Director of the Reshape & Innovation Center (@Reshape) at the Radboud University Nijmegen Medical Center (@umcn) in The Netherlands. His aim is to turn healthcare into an environment in which patients can act as partners; one of the ways to achieve this is with the help of innovations. He is faculty of the @Futuremedtech track of the Singularity University. He founded @MedCrowdFund™ & @MedCrowdPitch™, a crowdsource and crowdfund platform for patients to design healthcare-research; FaceTalk (@facetalkme), a videoconference system in which patients can join en.facetalk.nl); and The Future of Health conferences (@thefuturehealth) like TEDxMaastricht; and is mentor for several international health start-up programs. (@lucienengelen.)

Richard Fitton, MB, BS, MRCS LRCP, DCH, DRCOG MRCGP, is a General Practitioner in Manchester, England. He has been involved in patient records since 1968. He built a patient-centered medical centre in 1996 in which he researched patient accessing, contributing to and correcting their GP records. He has recently co-authored guidance for clinicians to share the complete and contemporaneous electronic record with patients. He has been a member of the National Care Record Development Board, the DOH Working Group on Copying Letters to patients, the All Party Parliamentary working group on "Technology in Medicine–Telehealth," the Bethesda, Maryland (U.S.) roundtable conference on Electronic Health Records, and The Wellcome Trust working group on the use of data for research. Dr. Fitton is a member of the International Society of Urban Health and has presented patient record access to the WHO Family of International Classification and to UNESCO in Paris.

Stasia Kahn, MD, is the President of Symphony Medical Group, an independent primary care medical practice in Illinois. Dr. Kahn's passion is to use health information technology to improve safety and quality of her patient's care. She is a champion of health information technology and quality healthcare, participating in practice-based research on Health Information Exchange and Quality Reporting from EMRs. She shares her practical experiences with others through EMRSurvival.com, using a blend of original articles, books, and audio casts. Dr. Kahn serves as an Advisor to the State of Illinois Health Information Exchange. She is a former member of the Greater Chicago HIMSS Board of Directors and former lecturer in the Masters Degree program in Medical Informatics at the School of Continuing Studies at Northwestern University.

Joseph C. Kvedar, MD, is Founder and Director of the Center for Connected Health, creating and validating connected health solutions that empower patients and providers to transform care. Dr. Kvedar is developing interventions that motivate behavior change, generate efficiencies and improve the quality of care. He is internationally recognized for his leadership, is a frequent lecturer and has authored over 70 publications on the subject. Dr. Kvedar is a board member for a number of organizations, including Continua Health Alliance and Care Continuum Alliance. He is a Co-Founder of Wellocracy, a community dedicated to empowering and engaging people to simply and effectively manage their health and wellness in ways never before possible, and a Co-Founder of Healthrageous and Chair of the company's Scientific Advisory Board.

DeeAnna Merz Nagel, LPC, DCC, BCC, is a psychotherapist and international expert regarding the impact of technology on the helping professions. She is co-founder of the Online Therapy Institute and Managing Co-Editor of *TILT Magazine ~ Therapeutic Innovations in Light of Technology.* Her presentations and publications include ethical considerations for the helping professions with regard to online counseling, social networking, mixed reality and virtual world environments. She has co-authored/edited two textbooks and written several book chapters and articles on topics related to technology and mental health. Ms. Nagel graduated from the University of Georgia with a M.Ed in Rehabilitation Counseling and is licensed to practice in New Jersey, New York and Georgia. She is a Certified Rehabilitation Counselor, Distance Credentialed Counselor and a Board Certified Coach.

W. Ryan Neuhofel, DO, MPH, called "Dr. Neu" by his patients, is a board-certified family physician practicing in Lawrence, Kansas. After completing residency at University of Kansas Medical Center in 2011, he started a solo "Direct Primary Care" practice, NeuCare Family Medicine. Dr. Neu provides a broad spectrum of primary care in a unique fashion that includes both old-fashioned and high-tech services. In between doing house calls, he uses the social media, email and webcam to connect with his patients. Dr. Neu also uses video and multimedia to educate on a variety of health topics. He envisions a future medical home where high-tech, affordable primary care is built upon a direct cooperative relationship between patient and doctor.

Gerry L. Tolbert, MD, is a board certified family physician and co-owner of a direct primary care practice in Northern Kentucky. He currently serves as a New Physician representative to the American Academy of Family Physicians Congress of Delegates and is the Medical Director for a free clinic in Falmouth, KY set to open in 2013. His interests lie primarily in patient communication technologies and improving EMR user interface.

Josh J. Umbehr, MD, is a co-founder of AtlasMD Concierge Family Practice. Dr. Umbehr (known as Dr. Josh) earned his Bachelors of Science in Human Nutrition at Kansas State University and then went on to earn his medical doctorate at the University of Kansas. Dr. Umbehr completed his residency in Family Medicine at Wesley Medical Center in Wichita Kansas and has been Board Certified since 2010. AtlasMD is Wichita's premiere concierge family practice and is an innovative solution to expensive insurance-based care.

Michael H. Zaroukian, MD, PhD, FACP, FHIMSS, is Chief Medical Information Officer for Sparrow Health System in mid-Michigan. He is also Professor of Medicine and former CMIO at Michigan State University. A practicing general internist, Dr. Zaroukian provides leadership to the implementation and optimization of Sparrow's EHR and associated information systems as care transformation-enabling technologies. Dr. Zaroukian was elected to the HIMSS Board for 2013-2016, received the 2010 HIMSS Physician IT Leadership Award, and chairs the HIMSS Ambulatory Information Systems Committee. Dr. Zaroukian also chairs the American College of Physicians Medical Informatics Committee, is a board member for the Association of Medical Directors of Information Systems (AMDIS), and serves on the American Medical Association (AMA) Health Information Technology Advisory Group.

About the Cover Artist

Regina Holliday is an artist, speaker and author in Washington, DC. She writes about the benefits of health IT and timely data access for patients and families. She painted a series of murals depicting the need for clarity and transparency in medical records. Her late husband, Frederick Allen Holliday II, inspired this advocacy mission. Regina paints and speaks at medical conferences throughout the world. She also began an advocacy movement called "The Walking Gallery." The Gallery consists of medical providers and advocates who wear patient story paintings on the backs of business suits. The painting that graces the cover of this book is entitled "Hope." It is the jacket image for Jan Oldenburg. This image encapsulates the hope that digital health is providing to patients everywhere.

Acknowledgments

Dave Chase's family supported the extra time he spent on this project, especially his wife, Coleen Chase, who is his editing partner. He also thanks his daughter Abby Chase and son Cam Chase.

Kate Christensen thanks her husband, Jim Rose, who edited her chapter and provided moral support.

Jan Oldenburg thanks her husband, Jon Ninnemann, and her sons, Andrew and Matt Ninnemann, who provided moral support and coped with her distraction during a summer focused on writing. She also thanks her sister, Julie Phend, and friends Susan Furlow, Judy Derman, and Laura Faulker, who reviewed and edited chapters, provided input on titles, and generally supported the project. Meg McCabe heroically read the entire manuscript and provided suggestions and perspective.

Brad Tritle's thanks and acknowledgements go to his wife, Ayako Tritle, and daughter, Aisha Tritle, who put up with nights and weekends at the local coffee shop, edited chapters and gave moral support.

The editorial team extends deep gratitude to those who gave of their time and expertise to write chapters for the book or gave permission to use blog materials posted elsewhere:

Bianca K. Chung, MPH
David Featherstonehaugh
Adam Greene, JD, MPH
Deven McGraw, JD, MPH, LLM
Jane Sarasohn-Kahn, MA, MHSA
Aaron E. Sklar, MS
Jonathan Wald, MD, MPH, FACMI

Special thanks are extended to artist Regina Holliday, who contributed the wonderful cover art, and to Daniel Z. Sands, MD, MPH, FACP, FACMI, who wrote the Foreword for this book.

We are deeply grateful to the people who generously shared stories and wisdom by writing case studies for the book:

Natasha Burgert, MD, FAAP
Nancy Burghart-Hall, CHCIO
Karen Colorafi, RN, MBA, BScN, CPEHR, CPHIT
Susannah Fox
Amir S. Hannan, MBChB, MRCGP, and Glen Griffiths, MSc, MBCS
Peter Hudson, MD
David W. Moen, MD
Eve Phillips, M.Eng, MBA
Sue Sutton, RN, PhD
Deborah C. Wells, MS, CPHIMS
Susan Woods, MD, MPH

We also deeply appreciate those who allowed us to interview them and shared their wisdom both for case studies and vignettes:

Kenneth Adler, MD, MMM
Michael Albainy
Sigal Bell, MD
Douglas Bonacum
Jen Brull, MD, FAAFP
Joseph Cafazzo, PhD, PEng
Steven Davidson, MD
Dave deBronkart
Lucien Engelen
Ted Epperly, MD, FAAFP
Ted Eytan, MD, MS, MPH
Richard Fitton, MB, BS, MRCS LRCP, DCH, DRCOG MRCGP
Matt Handley, MD
James Joun
Stasia Kahn, MD
Charles Kennedy, MD
Rob Klusman
Joseph C. Kvedar, MD
Howard Luks, MD
Paulo Machado
Diane McNally

W. Ryan Neuhofel, DO, MPH
Linda Stotsky
Wendy Sue Swanson, MD, MBE, FAAP
Paul Tang, MD
Gerry L. Tolbert, MD
Josh J. Umbehr, MD
Greg Weidner, MD
Michael H. Zaroukian, MD, PhD, FACP, FHIMSS

Finally, we thank those who supported us with discussion and ideas:

David Bowen, MD
Jeff Brandt
Steven Davidson, MD
James Joun
Mary Griskiewicz
Valerie Knoke
Paul Machado
Andy Oram (especially helpful in the writing of Chapter 6)
Helen Ovsepyan (invaluable contributions to Chapter 10)
Laila Taji
Greg Weidner, MD

We also want to thank all of the individuals who participated in crowdsourcing the title! If it takes a village to raise a child, this book is proof that it takes the wisdom of a crowd to make a book.

Contents

Foreword

By Daniel Z. Sands, MD, MPH, FACP, FACMI
Assistant Clinical Professor of Medicine, Harvard Medical School
Co-founder and Past President, Society for Participatory Medicine

Healthcare providers often erect barriers—both informational and logistical—that prevent patients from engaging in their care, which leads to healthcare that is costly, inefficient, of variable quality, and not satisfying to patients or providers. In the U.S., "Meaningful Use" regulations are trying to change this dynamic through financial incentives to providers, but this misses an important point: we must change not only how we practice, but also our attitudes toward the patient-provider partnership and the culture in which we practice. Doing so will create a better healthcare system for both patients and providers.

Francis Bacon wrote, "Knowledge is power." Consciously or unconsciously, we healthcare providers have used this as the basis for empowering ourselves and disempowering patients. We have been acculturated to a model of information asymmetry, in which physicians believe they know everything and that their patients are capable of knowing very little. Physicians thus become the all-knowing oracles, doling out information to patients, time permitting, and only on a need-to-know basis. Although this may feel empowering to providers, I'd suggest that it is a burden. After all, who among us actually knows everything—or feels as if we do?

Furthermore, this model is no longer realistic because the information landscape has changed. Health information is as available to laypeople as it is to healthcare professionals, including access to the world's medical literature. Susannah Fox and the Pew Internet and American Life Project surveys have taught us that, although health professionals are the preferred source of health information, 80% of online adults have gone online to look for health information, and people in all age groups (including the elderly) commonly engage in online health activity.[1] Over half of these health information seekers change their health behavior as a result.[2] In addition, PwC reports that one third of consumers are using social media for health discussions.[3] We healthcare professionals undervalue ourselves if we believe our value to be omniscience in our fields. In fact, our skill (beyond procedural expertise) is in knowing how to find and interpret information and to apply it to an individual patient, as well as to serve as a comforter and healer in the Hippocratic sense.

And what about medical records, once the sole purview of clinicians? In the U.S., patients have been guaranteed access to their medical records since the HIPAA Privacy Rule became effective in 2003, although not always without friction and burdensome costs (as Regina Holliday's tragic tale illustrates[4]). As medical records are increasingly stored in electronic format, access to those records can be facilitated.

Clearly, we're fooling ourselves if we think we have a lock on healthcare information access.

We must therefore embrace a model of information symmetry, in which the patient and clinician are partners, collaborating around the patient's health. In fact, the Institute of Medicine's 2012 report, "Best Care at Lower Cost,"[5] says patient-clinician partnerships ("engaged, empowered patients") are a cornerstone of creating the continuously learning health system that America needs.

In this model, information is freely exchanged because hoarding information bestows no power and runs counter to the common goal. How can patients be expected to take care of themselves if they don't have access to their own health information? This model will actually relieve physicians of the burden of pretending that they know everything, and be a more comfortable practice model for all.

Beyond just restricting access information, we have a healthcare system in which there are logistical barriers to receiving care. A major reason that patients do not schedule or reschedule timely appointments is that it's a frustrating experience. They must make a telephone call, wait on hold or await a call back, and then try to book an appointment into an overcrowded appointment schedule. Waits of one to two months are not uncommon. This is all the more upsetting because while we make people in need wait, we bring people into the office who don't need to be seen, including stable patients with hypertension, patients returning just to get their test results, and patients requesting prescription renewals. Why do we do this?

Contrast this with other industries. When was the last time you waited in line for a bank teller for a routine banking transaction? Or used a travel agent for a routine travel arrangement? Other industries have taken friction out of the system, which improves efficiency and optimizes resource utilization among all stakeholders.

Certainly some of this behavior is driven by a fee-for-service reimbursement system. Clearly, as we in the U.S. migrate our reimbursement system to one that pays for value and patient satisfaction rather than quantity of care provided, it will unleash provider creativity to deliver care in a more efficient and effective manner. But even today, how can one argue against making it easier for patients to schedule appointments? And as for e-communication, I have heard countless physicians in fee-for-service practices tell me that they won't use e-mail with patients unless they get paid for it. While that may be one approach, there are plenty of business reasons in a fee-for-service environment to manage some patient needs through asynchronous e-communication rather than in visits: increased patient satisfaction, which attracts and retains patients; improved visit efficiency; better time management and documentation compared with utilizing only the telephone to communicate with patients; and the flexibility to reserve office visits for higher complexity patients for which they are better reimbursed.[6] And this is before we consider the incentive payments for communicating with patients electronically that will be coming with Stages 2 and 3 of Meaningful Use.

Electronic tools, such as self-service portals, secure e-communication, mobile tracking tools, and perhaps other modalities might go a long way to improving how we service our customers (the patients).

As we introduce technology, we must be mindful of how we do it. For example, a computer in an examination room can either drive a wedge between the patient and clinician, or it can enhance the conversation. Part of this has to do with eye contact, and part with the arrangement of the screen. We must learn how to use this technology most effectively or it will detract rather than enhance the relationship.

So how should providers get started? How might they take their first tentative and awkward steps in this new dance of patient engagement? This should start with a change of attitude and behavior. Recognizing that many of our patients are getting healthcare information online, let's begin by asking *every* patient (not just those who are young, well-educated, and of high socio-economic status) whether they go online to look for health information. We will not only be surprised by those who tell us that they do, but may also be surprised by those who don't. Record this in their record, just as you record information about other health habits, like tobacco, alcohol, and drug use, occupational history, and use of complementary and alternative medicine. By doing so you give tacit approval of this activity, which will result in honest disclosures of such health-seeking information in the future. And this will ensure that there will be open dialog about information seeking that may impact the patient's health in the future.

Next we must admit to ourselves that we don't know everything. Once we've done that, we're ready for admitting that to others. The next time a patient asks a question in the office to which you don't know the answer, don't wave your hands, feign deafness, make up a response, or leave the office to find out. Just turn to the patient, look that patient straight in the eyes, and say, "I don't know." While this may run counter to all of your years of training and be quite difficult, it is actually quite liberating. Once you've gotten comfortable doing this, move on to the next step: say to the patient who asks the tough question, "I don't know; let's look it up together." Then take a minute or two to begin to answer their question. This will catch your patients off guard but they will appreciate your candor and will grow ever more loyal to you and your practice.

This evolving model of care—in which both patients and their clinicians are partners collaborating around the patient's health—has been called participatory medicine. The name evokes an idea that healthcare, to be most successful, must not be viewed as a spectator sport, in which the patient expects to be treated without being fully engaged and clinicians are complicit in this arrangement. Rather, healthcare must be a participatory sport, in which both parties are fully engaged. If we reimagine healthcare in this way, information sharing, e-communication, non-visit-based care, and open medical records will naturally follow. For example, it is easy to envision (especially after the OpenNotes study[7]) the medical record becoming a shared instrument of patient-provider collaboration. In 2009 I co-founded the Society for Participatory Medicine to promote these concepts in healthcare.[8]

Am I proposing creating a healthcare system that is entirely patient-centered without consideration for the provider side of the relationship? Clearly patients are overdue for experiencing a healthcare system built around them. As Dave deBronkart (a.k.a. e-Patient Dave) has pointed out,[9] healthcare is the only industry in which the customers (patients) are generally not treated like customers. But in this case, practicing differently benefits clinicians, practices, and health systems as well as patients.

In this book, Jan Oldenburg and colleagues present a wealth of information for healthcare organizations trying to make this vision become a reality through the application of tools and technologies that can help engage consumers in their health and healthcare. Through discussion, literature review, examples, and concrete suggestions, readers can help transform their organizations into the patient-centered institutions they must become if we are to attain a high-performing healthcare system.

REFERENCES

1. Fox S. Health information is a popular pursuit online. Pew Internet & American Life Project. February 2, 2011. http://pewinternet.org/Reports/2011/HealthTopics/Part-1/59-of-adults.aspx. Accessed January 6, 2013.

2. Fox S. The social life of health information. Pew Internet & American Life Project. June 11, 2009. http://www.pewinternet.org/Reports/2009/8-The-Social-Life-of-Health-Information/01-Summary-of-Findings.aspx. Accessed January 7, 2013.

3. PwC Health Research Institute. Social media "likes" healthcare: from marketing to social business. April 2012.

4. Holliday R. Thoughts on medicine and social media. Regina Holliday's Medical Advocacy Blog. October 23, 2009, at http://reginaholliday.blogspot.com/2009/10/thoughts-on-medicine-and-social-media.html. Accessed December 23, 2012.

5. Institute of Medicine. Best Care at Lower Cost: The Path to Continuously Learning Health Care in America. http://www.iom.edu/Reports/2012/Best-Care-at-Lower-Cost-The-Path-to-Continuously-Learning-Health-Care-in-America.aspx. Accessed December 23, 2012.

6. Delbanco T, Sands DZ. Electrons in flight—e-mail between doctors and patients. *N Engl J Med.* 2004; 350:1705-1707.

7. OpenNotes. http://www.myopennotes.org. Accessed December 23, 2012.

8. The Society for Participatory Medicine. http://www.participatorymedicine.org. Accessed December 23, 2012.

9. deBronkart D. Would your doctor pay for wasted time? http://e-patients.net/archives/2011/06/would-your-doctor-pay-for-wasted-time-cnn-com.html. Accessed December 23, 2012.

Preface

By Jan Oldenburg, FHIMSS

Many aspects of healthcare delivery and financing are broken today. Patients are disconnected from their symptoms and their clinical data, and doctors spend too much of their time on frustrating administrivia rather than patient care. There are pockets of exceptional and innovative care in the United States and elsewhere, but in general, the healthcare system in the U.S. gets failing grades on the cost of care, the outcomes produced, the services rendered and the overall patient experience. Health plans and employers try to focus consumer attention on wellness activities and responsible healthcare spending, but they are disconnected from most consumers' experience of visiting their doctor or dealing with an illness. Though patient care is the core purpose of the healthcare system, that system has treated patients more like widgets than like individuals who have knowledge, skills, preferences and capabilities in their own right.

During the past decade, there has been a growing realization that for healthcare to be fixed, patients need to be equipped, enabled, empowered and engaged, working in partnership with providers to change the system. And to be able to function as true partners in their own care, patients need access to their data and a new relationship with their doctors founded on coming together in a spirit of mutual inquiry and learning. Advances in technology and the increasing use of electronic medical records enable use of personal health IT tools to empower and engage patients.

Discussions of patient engagement tend to focus on patients and how to motivate them to change—and it is an important factor. The discussion, however, tends to ignore the toll the current system exacts on providers. Providers bear the stress of needing to be infallible, of taking personal time to keep up with current literature on obscure conditions, and of making the right decision every time, and in a way that is sensitive to the patient's personal, clinical and financial situation.

We wrote this book to focus on the benefits of patient engagement, but from the standpoint of doctors, providers, and others in the healthcare system. It explores what is working in this new, digitally-empowered collaborative environment, using research and case studies to explore the benefits to individual providers of practicing in this way as well as the benefits to their organizations and staff. There are other works that use the patient as the frame of reference, such as *Doctor, Your Patient Will See You Now* by Steve Kussin or *e-Patients Live Longer* by Nancy B. Finn, M.Ed.

Main chapters of the book focus on the changing landscape of patient engagement, starting with the impact of new payment models and Meaningful Use requirements in Chapters 2 and 3. We review the impact of patient engagement in key areas of healthcare: patient safety, quality and outcomes, effective communications, and self-service transactions in Chapters 4 through 7. Chapters 8 and 9 explore the impact of social media and mobile as tools for engaging patients and changing healthcare. Chapter 10 provides guidance about privacy and security challenges, while Chapter 11 shows how designing in patient-centered ways can be incorporated across the system. Chapter 12 assesses the future direction of patient engagement.The main section of the book ends with advice about how to get started in Chapter 13. Interviews and vignettes throughout the book bring the topics to life, and the 23 case studies at the end of the book showcase the impact in a wide variety of settings, from large providers to small practices, from traditional medical clinics to eTherapy practices, as well as presenting several patient stories.

This book grew out of the work of the HIMSS eConnecting with Consumers Committee, and out of its predecessor, the Personal Health IT Task Force. In April of 2012, David Rowe, then Chair of the Personal Health IT Task Force, worked with Mary Griskewicz, Senior Director of Healthcare Information Systems for HIMSS, on a proposal to the HIMSS Board to elevate the Personal Health IT Task Force to a formal HIMSS committee, with a book on patient engagement one of its projects. The HIMSS committee and the workgroups it oversees have created a nexus for discussion of patient engagement values and strategies across organizational boundaries. The discussion supports the HIMSS' pledge to "equip clinicians and front line personnel with the necessary health IT tools to engage patients and consumers in healthcare. This is accomplished by providing educational programs, resources and tools and thought leadership with our members for the healthcare industry."

Most of the contributing editors and many of the chapter writers have collaborated with one another in the last several years in the context of HIMSS and the task force/committee work, as well as through shared experiences at conferences, through publications, and through the ongoing dialogues in forums like Twitter. This book represents an opportunity to tell the story of the way patient engagement is potentiated by technical change and the growth in IT capabilities.

The editorial team for this book has a longtime focus on the use of health IT tools and patient engagement and a deep commitment to this work.

In the 1980s, Dr. Kate Christensen was exploring how to make her clinical work life more efficient by experimenting with her own templates and databases to figure out how to serve patients better and more efficiently. As the Medical Director for kp.org during the last 12 years, as Kaiser Permanente implemented its clinical Personal Health Record, My Health Manager, she has seen the enormous impact of these tools in transforming clinical practice and provider-patient interactions across 9 million members. As the Chair of the HIMSS eConnecting for Consumer's Committee, she is in a unique position to discuss the impact of technologies for patient engagement on physician practices through contributing to this book.

Brad Tritle's experience in running a statewide HIE/HIT organization also led to a passion for this work. He explained the impact by saying, "I had previously spent much of my career in telecommunications and e-government. When I first began to work in health IT, I was astounded by the general lack of interest in engaging patients with information, as consumers were already successfully interacting with their financial and educational institutions, and their electric and telecom utilities and government agencies electronically. I saw a great disconnect: an industry begging patients to take more responsibility on one hand, while on the other hand refusing to provide them with the information and tools that would enable them to take that responsibility. When I was asked to design and facilitate consumer focus groups around health IT, including HIE and PHRs, security and privacy, it further and deeply impacted me. I heard firsthand about consumers' unmet needs and their desire for a system that would serve and respect them, rather than leave them feeling as if they are herded like cattle (though some providers rightly say that some consumers prefer to be told what to do). My desire to build such a system, in the hope that it would also positively impact quality, costs, and population health, continues to fuel my passion for consumer engagement."

Dave Chase notes that being a part of the "sandwich generation" has given him a three-generational view of the healthcare system. In addition, he has supported several friends through the cancer journey. Those experiences made it clear to him that as an industry, we could do much better at delivering healthcare—and also that there is a disconnect in that we are moving to accountable care models, yet most processes and workflows ignore the most important members of the care team: patients and their families or caregivers. He said, "Remember, only 1% of a person's life is spent with healthcare professionals. Further, 75%-80% of healthcare spending is related to chronic disease and it's clear that the vast majority of decisions that have the *most* impact on outcomes are made by patients and families, not by professionals. Decisions such as whether or not to fill and use prescriptions, what to eat, whether to exercise—those are made by patients and family members or caregivers. I'd spent considerable time in both enterprise and consumer software markets, and knew there was a way to combine the best of both worlds to make a big difference in how healthcare is delivered and how patients are engaged."

Dr. Christensen asked me to lead the work on this book during my last week at Kaiser Permanente, during our "farewell lunch," just after I had accepted a new role within Aetna's Accountable Care Solutions team as the Vice President for patient and physician engagement. Like Kate, my passion for this work was fueled by seeing the transformative effect of patient engagement using personal health IT tools at Kaiser Permanente. My journey has also been a personal one, though I rarely talk about that aspect, fearing that I will lose credibility if I am seen as "patient" rather than "professional." In my early 30s, I was diagnosed with asthma, an unusual consequence of pregnancy. For several years, asthma ruled my life. My disease didn't present in typical ways and it struck at the core of how I saw myself: articulate, independent, competent, strong. In an effort to understand what my triggers were, I began tracking my peak flows. I recorded peak flow readings three times a day. I graphed them on Excel. I overlaid when I was on prednisone and at what levels. I took those graphs with me to urgent care and doctor's offices. It was partly defensive: if I couldn't get the words out because I was coughing, I wanted my data to speak for me. Several of my doctors loved having data and worked closely with me to understand and manage my condition: my pulmonologist, Dr. Ieva Grundmanis, and my ENT, Dr. Thomas Christiansen. More often, though, doctors responded with wariness and suspicion, more apt to label my graphs and charts as more indicative of obsession than an effort to participate with them in understanding and managing my condition. My tracking taught me a number of key lessons about my disease, but it also left me with an enduring passion to help other patients to empower themselves with knowledge and data.

Consumer demand still far outstrips the capabilities offered in healthcare. Despite the fact that 65% of patients want access to their records and 53% of patients want email access to their providers, as described in a recent Harris Interactive poll, only 17% of patients report having access to their records; only 12% can email their doctors, and only 11% report being able to manage appointments online.[1] Many providers and health systems are still wary of introducing these capabilities, concerned about the impact on the way they practice and concerned that patients won't be able to handle technical clinical data. We wrote this book to alleviate those fears: to gather all the research we could find to build the case for patient engagement and describe why it is not only good for patients, but also for providers and the healthcare system as a whole.

Our goal is for this book to enhance the dialog within and across healthcare organizations, as well as between providers and patients. In patient engagement, we see the hope for transformation in the way patients, families, and providers work together to live happier and healthier lives.

This book would not have come to life without the volunteer commitments of the editors noted above, as well as the chapter contributors, case study writers and interviewees who generously gave their time, energy, and thought leadership to this work. It also would not have come to being without the energy and enthusiasm of Mary Griskewicz from HIMSS, as well as the HIMSS editorial staff, Nancy Vitucci and Fran Perveiler. Regina Holliday contributed the cover art, "Hope"—created as a jacket for the walking gallery—with its depiction of hope and of patients and providers using technology to work in partnership. Dr. Danny Sands graciously wrote the Foreword. The book also incorporates the work of many other thinkers in the field who have contributed research, thought leadership, and support to building a new healthcare system that will engage both patients and providers.

We hope you enjoy this book, and that it stimulates a different kind of conversation inside your organization about patient engagement: not whether to do it, but when and how.

Endnote

1. Harris Interactive. Patient choice; an increasingly important factor in the age of the healthcare consumer. September 10, 2012. http://www.harrisinteractive.com/NewsRoom/HarrisPolls/tabid/447/ctl/ReadCustom%20Default/mid/1508/ArticleId/1074/Default.aspx. Accessed October 14, 2012.

Chapter 1

Transforming Healthcare Through Digital Patient Engagement

By Jan Oldenburg, FHIMSS

INTRODUCTION

Imagine a future where patients' daily interactions with the healthcare system are simple, straightforward, and enabled by personal health information technology (IT) that is available whenever they want and wherever they are. Where patients have full access to their clinical and financial data—across time and across providers—and can use it both to educate themselves about their health and to ensure that their medical records are correct. Where patients have devices that provide moment-by-moment feedback to them (and, in the right context, their care team) about the effect their actions are having on their health in real time—rather like the way Prius drivers can tell whether the way they are driving uses more or less gas. A future where patients can take care of health tasks such as scheduling appointments or paying bills whenever it is convenient for them. Where they can enlist family and friends in supporting them, whether they are trying to recover from a hospital stay, provide an update on their medical condition, or include more exercise in their daily routine. And where, when they fall ill, they can seamlessly communicate with their doctors, care team, and hospital. In this imagined future, patients become a key part of the care team, actively participating in decisions about what will work for them and their health—physical, emotional, and financial.

Imagine that when personal health IT tools are used by patients as described above, providers are able to eliminate non-productive parts of their current work and focus more deeply on the things that motivated them to go into healthcare in the first place—spending time with patients—helping them when they are sick, coaching them to get and stay healthy. Now precious clinic resources can be used where they are needed most, because self-service programs allow patients to conveniently manage tasks themselves. Providers can keep track of their patients and manage minor problems through virtual appointments or email exchanges. In this imagined future, providers collaborate with patients to develop plans that reflect the totality of the patient's health: patients are more satisfied and providers are happier, feel less rushed, and feel less pressure to be perfect.

For the fortunate few patients and providers, aspects of this future are here today. In the health systems in which personal health IT tools are available to patients, the dynamic between patients and providers is evolving, personal health IT tools are helping to drive costs from the system, and satisfaction levels for patients are rising.

For too many patients and doctors, however, this is a distant reality. Many studies (referenced elsewhere in this book) show that patients desire capabilities such as online access to their records, the ability to exchange secure messages with their doctors, and the ability to manage appointments online. A Harris Interactive poll from September 2012 highlighted the fact that despite those wishes, only 17% of patients report having access to their records, only 12% can email their doctors, and only 11% report being able to manage appointments online. The complete numbers are highlighted in Figure 1-1.[1]

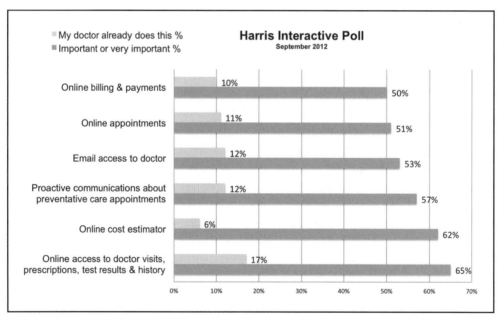

Figure 1-1. Patient Desire vs Availability of Personal Health IT Tools

This book explores the way personal health IT tools that engage patients also benefit providers and the healthcare system as a whole. The case for patient engagement is a key component of the case for healthcare transformation.

WHAT IS PATIENT ENGAGEMENT?

For all the conversation about patient engagement, the healthcare industry hasn't agreed on what the phrase really means. A survey conducted in June 2012 by the National eHealth Collaborative (NeHC) on consumer engagement with health IT found a wide range of definitions of patient engagement, as well as a wide range of goals among partner organizations.[2]

The Center for Advancing Health Behavior offered a good framework for patient engagement in general by defining engagement as[3]:

"Actions individuals must take to obtain the greatest benefit from the healthcare services available to them."

The Center went on to further explain their definition in ways that we wholeheartedly endorse:

"This definition focuses on behaviors of individuals relative to their healthcare that are critical and proximal to health outcomes, rather than the actions of professionals or policies of institutions.

Engagement is not synonymous with compliance. Compliance means an individual obeys a directive from a healthcare provider. Engagement signifies that a person is involved in a process through which he or she harmonizes robust information and professional advice with his or her own needs, preferences and abilities in order to prevent, manage and cure disease."

Our definition of patient engagement using personal health IT tools expands on the Center's definition: Patient engagement using personal health IT tools comprises actions individuals take to obtain the greatest benefit from the health services available to them, using any information technology capability that enables them to:

- Better understand their health and health conditions.
- Obtain access to their own health data in real time or near real time.
- Improve communications with their doctors and providers.
- Take more responsibility for their own health and health outcomes.
- Improve their experience of interacting with the healthcare system.
- Inform and educate their families and caregivers.
- Get support about health and healthy behaviors from family, friends, caregivers, and health professionals.

Capabilities that impact these areas may be offered by a physician, health system or payer, or they may exist as stand-alone tools offered independently of any health system entity. The definition includes mobile and web-based capabilities so that individuals can choose the platform and channel appropriate for them. The question of who *owns* the data is irrelevant as long as patient data are stored securely, patients have access to their own data in near real time and can choose to download or transmit it, and information about the source of the data is maintained.

Engagement and Empowerment vs. Compliance

When people in the healthcare industry talk about patient engagement, they frequently refer to patient compliance–behaviors they want their patients to do "for their own good." They want people to exercise more, eat healthier, understand their diseases, take their medicines, and shop for healthcare in a more informed way. By this definition, patient engagement is the Holy Grail of healthcare. If consumers and patients would only consistently behave in ways that are good for them, many of the preventable causes of disease would be eliminated or at least postponed.

As noted in the definition cited earlier, however, engagement is not synonymous with compliance. Many of us in healthcare are guilty of wanting compliant patients rather than engaged, empowered or self-actualized patients because compliant patients are more likely to follow instructions than to ask questions or challenge instructions. Being challenged can be uncomfortable, but for healthcare transformation to occur, it is a discomfort that healthcare professionals may need to embrace for the good of their practice and healthcare overall (see Figure 1-2).

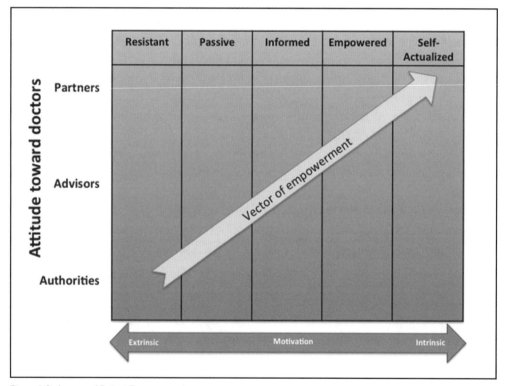

Figure 1-2. Journey of Patient Empowerment

There are many stages of engagement, as depicted in Figure 1-2. People choose ways to engage that feel right and fit their needs, preferences, and personal abilities. Patient engagement offers the opportunity for patients to move toward empowered and self-actualized behaviors. With those shifts, there is also a logical progression from viewing providers as authorities to viewing and working with them as partners.

Providing the opportunity for patients to move toward empowered and self-actualized behaviors does not guarantee that they will behave in healthy ways.

There is no single answer to the question of what motivates people to change or why some people succeed and some fail, much less how to sustain behavior change over time. Providing incentives and external motivation can help motivate change, but for sustained change to occur, external motivation needs to be replaced by internal motivation. Patients do not need to become fully self-actualized to see a positive impact on their health, but one important way we can support individuals' movement toward empowerment is by offering personal health IT tools that support them to the degree they are ready to interact. These tools help us engineer teachable moments that promote behavior change in patients based on delivering information, education, and a call to action at the moment each individual is most engaged in his/her health and most open to change.[4]

Why Now?

It has long been conventional wisdom that people play a role in managing their own health. Mark Twain once said, "The way to keep your health is to eat what you don't want, drink what you don't like, and do what you'd rather not." Benjamin Franklin reflected the same sentiment when he said, "To lengthen your life, lessen your meals." Alas, for the same amount of time, people have been ignoring the conventional wisdom and doing what they like, regardless of the impact on their health.

So what is different now? Why do we believe the time is right for patient engagement to come to the forefront?

We are at a historic inflection point where a number of forces are coming together to enable—and perhaps require—doctors and patients to engage with one another differently.

One important dimension of the change is the rising cost of healthcare. The current costs and cost trends are unsupportable. The cost of healthcare in the United States, combined with the relatively low quality of the care provided, is a key driver of change. It has led to healthcare reform, experimentation with new models of practice and compensation, and a renewed energy about changing the interactions between patients and their healthcare providers.

Another major factor is the digitization of health data and the rise of the Internet, which make it possible for patients to see and understand their health records and health history online and on mobile devices. Consumer movements such as the Society for Participatory Medicine (S4PM) are bringing attention to the rights patients have in relation to their healthcare data and the obligations providers have for being transparent about healthcare data. As patients become empowered to see their health records online, providers' attitudes toward the record must undergo a corresponding shift. Gone are the days when the doctor or nurse walked into the room, clutching your chart tight to the chest, carefully protecting you from seeing anything it contained. This shift, from a record that is *owned by* the provider to a record that is *shared with* or even *managed by* the patient, is the most visible manifestation of a significant change in the dynamic between patients and providers.

The rising tide of mobile devices and innovations that allow patients to complete healthcare transactions, track their health and exercise, and review their records wherever and whenever it is convenient supports this shift. By fostering innovation, organizations like Health 2.0, Rock Health, and even the Office of the National Coordinator (ONC) are encouraging these changes. Patients themselves are changing the relationship with their doctors and nurses, whether by bringing in Internet articles, tracking their health rigorously between visits, or just asking more questions about recommended treatments or the cost of care.

Other dynamics are changing for providers as well. The HITECH (Health Information Technology for Economic and Clinical Health) Act incentives are changing medical practice in hospitals and clinics by supporting and enabling investments in electronic health records (EHRs). CMS (Centers for Medicare and Medicaid Services), enabled by healthcare reform, is supporting and creating pilot projects—with incentives—of new forms of practice, ranging from Accountable Care Organizations (ACOs) to Patient-Centered Medical Homes (PCMH), Direct Primary Care (DPC), and Medicare Shared Savings Program (MSSP) arrangements. These arrangements shift incentives away from fee-for-service medicine and toward value-based care and encourage a focus on patient satisfaction. Employers and payers are a part of this change, which they can impact by creating a culture of wellness within the workplace, as well as by selecting health benefit designs that both encourage healthy behaviors and require patients to bear a significant share of the cost of treatment.

All of these forces intersect at this moment in time, changing—we think forever—the landscape and the environment for patient and provider interactions, offering both opportunities and challenges. Ted Eytan, MD, encapsulated this shift best when he said in an interview with the author:

> "I consider this to be the 'decade of the patient,' which to me means patients will be more involved and more central to what we do in healthcare than ever before. This includes not just access to the information in their medical records, but in the design and operation of the health system itself. Until the day comes when medical professionals always get it right, patient and patient advocates would like to help out, fill in the gaps, save themselves, the people they love, and their society from unnecessary harm. And lots and lots of physicians are going to stand with them."[5]

WHERE TO START

It is good to listen to what patients say they want when putting a patient engagement program in place. Figure 1-3 highlights key base-building components in place—basics that satisfy key patient needs.

A more detailed framework for patient engagement capabilities is included in Appendix A, and more ideas for getting started are provided in Chapter 13, but the key areas to pay attention to include providing patients with their clinical data and relevant education that empowers them; enabling convenient access to their physicians and care team with communication tools; providing tools that allow them to set goals and track changes over time; and self-service transactions that make it more convenient to interact with the healthcare system. Patient engagement is a lot about supporting and strengthening the secure, trusted relationships that already exist between patients and providers. These tools build a foundation that can simplify interactions with the healthcare system and support patient empowerment.

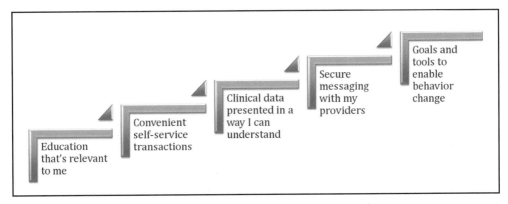

Figure 1-3. Key Patient Needs for Patient Satisfaction

Those of us who work in healthcare can provide capabilities that support people in their life journey. We can design or implement personal health IT tools that are simple and intuitive for patients—and their providers—to use. We can provide an array of capabilities that engage patients by educating them, entertaining them, and helping them find internal motivation. We can make it easier for patients to act in healthy ways, to ask questions, and to interact with the healthcare system. We cannot magically transform patients from resistant to self-actualized, but we can deploy personal health IT tools that support them to take steps along the way—and all of those steps, together, can move our industry toward transformation.

CONCLUSION

This is a book about the things that can and are being done today using personal health IT tools to make changes in the interactions between patients and providers, as well as patients and healthcare systems.

This book explores the ways we can use technology to help patients help themselves—and the way that doing so benefits not only patients, but providers and hospitals as well.

Why is patient engagement using personal health IT tools good for patients?

Patient engagement using personal health IT tools is good for patients because it:

- Enables them to better understand their own health, facilitating more-informed decisions.
- Provides them with convenient ways of accomplishing their healthcare tasks.
- Provides them with ways to interact with their providers.
- Provides them with ways to track and understand their health and health conditions.

Why is patient engagement good for providers?

Patient engagement using personal health IT tools is good for providers because it:

- Helps them to keep their patients healthy by educating and informing them.
- Helps them treat patients effectively when they are sick, using the right level of care.
- Builds patient loyalty.
- Reduces administrative costs.

Why is patient engagement good for the system overall?

Patient engagement is good for the healthcare system because it:

- Lowers cost in the healthcare system by reducing waste and redundancy.
- Satisfies providers and patients alike.
- Enables necessary change across the system as a whole.

REFERENCES

1. Harris Interactive. Patient choice; an increasingly important factor in the age of the healthcare consumer. Posted September 10, 2012. http://www.harrisinteractive.com/NewsRoom/HarrisPolls/tabid/447/ctl/ReadCustom%20Default/mid/1508/ArticleId/1074/Default.aspx. Accessed October 14, 2012.

2. National eHealth Collaborative. Consumer engagement with health information technology. Summary of NeHC Survey Results. Posted June 2012. www.nationalehealth.org/ckfinder/userfiles/files/Consumer Con.Meeting/CE_HIT_Summary3.pdf. Accessed October 14, 2012.

3. Center for Advancing Health. Patient engagement behavior framework: What is "Patient Engagement?" http://www.cfah.org/pdfs/CFAH_Engagement_Behavior_Framework_current.pdf. Accessed September 30, 2012.

4. Lawson PJ. Teachable moments for health behavior change: A concept analysis. *Patient Education and Counseling.* July 2009; 76(1):25-30. http://www.pec-journal.com/article/S0738-3991%2808%2900583-1/abstract. Accessed November 10, 2012.

5. Ted Eytan, MD, MPH; interviewed by Jan Oldenburg, September 30, 2012.

Chapter 2

Patients and Personal Health IT in the Era of Accountable Care

By Jane Sarasohn-Kahn, MA, MHSA

INTRODUCTION

To slow the growth of national health spending in the U.S., healthcare payers are moving from volume-based to value-based payment regimes. The objective of this shift from volume to value is to improve quality and reduce cost: in short, "no outcome, no income." Among value-based payment approaches are bundled payments, pay-for-performance, Patient-Centered Medical Home (PCMH) models and Accountable Care Organizations (ACOs). This chapter will focus on ACOs in order to explore more deeply the ways that these new models of care bring patient engagement and personal health IT tools to the forefront.

ACOs are a prominent feature in the Affordable Care Act of 2012 (ACA), established as part of the Medicare Shared Savings Program (MSSP). A Medicare ACO is accountable for delivering all healthcare services for an attributed group of patients, across the continuum of care—inside and outside of hospitals.[1] ACOs are motivated to achieve specific targets tied to patients' costs, outcomes, and satisfaction.

The ACO model has three core features:

1. Health provider organizations with a strong base of primary care, together which are "accountable" for quality and costs across the full continuum of care for patients enrolled in the ACO.
2. Performance measurement to support continuous improvement in care and population health outcomes.
3. Payments linked to quality and reduced costs.[2]

Thus, health providers in ACOs have strong financial incentives to work in clinical teams and to coordinate care across practice settings: inpatient, outpatient, community-based ambulatory care, and directly to patients. For health systems to fully reach their ACO measurement goals, they must reduce the cost of chronic illness. Getting patients onto the care team bolsters engagement and drives positive health outcomes.

To achieve these objectives, an ACO must build an IT infrastructure that enables healthcare communication, coordination, and collaboration across clinical team members, as well as with patients who must effectively engage in their own self-care processes.

This chapter discusses the importance of personal health IT to enable patient engagement for ACOs, the most relevant types of personal health IT tools for ACO patient engagement, and why this matters for providers, the health system at large, and most importantly, patients.

WHY PERSONAL HEALTH IT FOR ACOS?

More than one in four Americans is affected by multiple chronic conditions (MCC). Among Americans aged 65 and older, three out of four have MCCs.[3] According to the Centers for Disease Control and Prevention, three in four healthcare dollars spent in the U.S. go to managing chronic disease, such as asthma, diabetes, and heart disease—conditions largely amenable to patient self-care through lifestyle modifications such as those around food intake, physical activity, alcohol consumption, and smoking cessation, as well as adherence to prescribed medications.[4]

For ACOs, managing chronic conditions for patient populations is a critical factor for long-term financial viability. Managing chronic disease requires ongoing observation, communication between patients and providers, and recurring treatment adjustments. Most patients cared for in the U.S. health system are treated in an episodic, fragmented mode, compromising continuity of care and streamlined communication. Beyond their primary care providers, patients with chronic disease also visit specialists and other health professionals in the community who do not function as a coordinated healthcare team, further jeopardizing continuity of care and safe hand-offs, which can lead to higher costs and complication rates.[5]

"Chronic disease drives cost and to control it, you must get the chronically ill patients involved in managing their own diseases,"[6] advised Charles Kennedy, MD, CEO of Aetna Accountable Care Solutions. Patient-centered IT tools help to improve outcomes by improving communication, access to data where and when required, and patient self-care, based on a meta-analysis conducted by researchers from Johns Hopkins.[5] The team found that, despite numerous barriers to using health IT to enable patient-centered care, there is substantial evidence confirming that patient-facing health IT has a positive impact on health outcomes for a broad range of conditions such as cancer, diabetes, heart disease, and others.

Patients' use of information technologies can also effectively address lifestyle behavior change—in particular, when patients use more than one tool.[6] For example, MyPreventiveCare, developed at Virginia Commonwealth University, combined an electronic health record (EHR) with a personal health record (PHR) and patient alerts, driving positive health outcomes among a population of 30,000 patients cared for at 14 provider sites in the Virginia Ambulatory Care Outcomes Research Network (ACORN).

PATIENT-FACING HEALTH IT FOR ACOs

The Agency for Healthcare Research and Quality (AHRQ) assembled a useful list of patient-facing health IT tools in its omnibus report, *Enabling Patient-Centered Care through Health Information Technology*.[7] Table 2-1 organizes these tools by category.

Table 2-1. Categories of Personal Health IT (PHIT) Relevant for ACOs*

Categories of PHIT	Examples of tools
Secure electronic messaging	Email Social networking via peer-to-peer sites
Care management tools	Information technology-guided self-care Social networking Peer-to-peer support Health games
Telehealth	Remote health monitoring Video consultations with clinicians (remote visits) Mobile health applications
Personal health record and patient portal applications	Personal health record Patient portal
Shared decision making	Collaborative platforms for providers and patients

Source: Adapted from AHRQ. Enabling Patient-Centered Care through Health Information Technology, June 2012.

Selected personal health IT tools are discussed next, with their implications for ACOs.

Secure Electronic Messaging

In 2012, 39% of physicians communicated with patients through all digital methods, including 32% who communicated via email.[8] On the patient demand side, though, many more patients would like to exchange email with their physicians: 76% of consumers want to email their physicians, and 23% said they would pay extra for the opportunity to email their physicians.[9]

Email between physicians and patients is part of the design of Eisenhower Primary Care 365, a group practice in Rancho Mirage, CA, which embraces patient engagement for self-care. "Our first tier is online care, and patient visits become a secondary activity," said Joseph Scherger, MD, Vice President, Primary Care and Academic Affairs at Eisenhower Medical Center. In this medical model, care is primarily managed by patients who are supported by access to EHRs and online health education.[10] "It's the same transformation as money management, where you used to have a broker do the trading, but now they're just an advisor and you do the actual trading," Dr. Scherger said.[11] He described a recent vacation, during which he took care of his patients remotely with a few minutes of messaging a day except for one patient who he phoned from Maui; she was so delighted with his level of "patient service" that she recommended new patients to the medical practice.

Beyond customer service, physician-patient email has been found to bolster quality of care at Kaiser Permanente and other providers that have tracked the impact of emailing on patient care and physician workflow.[12] Ronald Dixon, Director of the Virtual Practice Project at Massachusetts General Hospital, demonstrated that asynchronous communication between patients and providers, based on Internet-based technologies like email and secure messaging, can help patients effectively manage conditions remotely.[13]

Care Management Tools

Researchers at Geisinger, one of the nation's "most wired" health systems,[14] developed the Technologies in Diabetes Education (TIDE) program to engage and motivate patients at risk for diabetes. The website engages patients through online education tools and games, and patients receive email tips on cooking and daily living. Patients enrolled in TIDE improved their self-care behaviors and experienced greater decreases in HbA1c than those in the control group. This program was particularly useful in Geisinger's target market, which includes rural areas.[15]

Another challenging aspect of care management is patient adherence to prescription drug medications. Two-thirds of adults do not take a medication as prescribed.[16] Tools for patients to aid in medication adherence are available, from mobile phone app reminders to pill bottles that glow when it's time to take a medicine. A new generation of ingestible pills was approved by the U.S. Food and Drug Administration (FDA) in July 2012, powered by contact with stomach fluid and working through a transmission signal that determines the identity and timing of a drug's ingestion to a patch worn on the skin. The patch sends the information to a mobile app, which can be accessed by clinicians and caregivers.[17]

Telehealth

Telehealth leverages the information and communications infrastructure to deliver healthcare remotely—between providers working from one location to patients located in another. Telehealth is moving out of medical centers and onto mobile devices, making it more accessible to a broader population and more types of care.[18]

Remote health monitoring, one application of telehealth, falls into four categories that are useful for providers in ACOs:

- Wellness and prevention, to keep patients' health in check, avoiding unnecessary visits to the doctor or emergency department, and preventing the onset of chronic disease.

- Chronic disease management, to enable patients to effectively manage their own care, and monitor conditions to avoid unnecessary visits to the doctor's office and exacerbations.

- Acute care, post-acute care, and rehabilitation, to ensure against re-admission to the hospital.
- Safe, healthy aging at home, to bolster seniors' wellness, self-care efficacy, and security.[19]

Conditions that have the greatest potential for remote health management include asthma, cancer, coronary artery disease, chronic heart failure, chronic obstructive pulmonary disease, chronic pain, depression, diabetes, and hypertension. A growing array of consumer-facing medical devices are available on the market that have USB-compatibility and/or operate wirelessly through the cloud, such as digital blood glucometers, blood pressure cuffs, WiFi-enabled weight scales, medication adherence tools, and body sensors that measure metabolic function.

Measurements from these devices can be communicated back to providers' offices for real-time monitoring, allowing clinicians to adjust patients' medications and therapeutic regimens as required.

mHealth Apps

Mobile health (mHealth) enables patients to manage conditions and wellness on a 24x7 basis, whether at home, at work, or during leisure time. mHealth applications are especially promising for people diagnosed with chronic conditions who must make daily choices about food intake and physical activity, as well as clinical decisions. Among the most complex conditions to manage is diabetes.

The Joslin Diabetes Center launched a mobile initiative, Joslin Everywhere, in February 2012, connecting patients with physicians. The online platform enables patients to digitally monitor blood glucose readings and communicate the data to clinicians, who can track patients' conditions and adjust therapies (e.g., medication, insulin dosing) as required. Joslin's vice president of planning told *The Boston Globe* that the medical center must migrate to remote health solutions because, "To be able to reach this tidal wave of people with diabetes coming at us, we need to use technology."[20]

In addition to the technology platform, Joslin Everywhere also offers patient support groups, remote weight management programs, and mHealth apps that communicate glucose readings to providers. Connected physicians in the program receive ongoing research, access to continuing medical education, and instructions on using the EHR system to identify candidates for the program.

In addition to clinical care support, patients can use mobile tools for administrative tasks in healthcare such as initiating electronic prescriptions and making appointments online. More hospitals are contracting with mobile app developers to provide patients with tools that help them compare emergency department wait times, calculate directional maps, and access appointment schedules for physicians online. These services give consumers the convenience they seek in a retail health mode and drive greater patient satisfaction.

Personal Health Record and Patient Portal Applications

Patient portals can provide customized health education to newly-diagnosed patients and those caring for chronic conditions, as well as providing patients with direct access to clinical data about them, such as laboratory test results and current medication lists. In addition, portals can include health-tracking tools that allow patients to record health measurements that can be observed and analyzed by clinicians. Health trackers can enable patients to manually record their data, but increasingly, they can use connected devices that feed data such as glucose levels and blood pressure wirelessly into databases that simplify tracking and analysis. Consumers look for portals to be accessible from mobile phones—once again, a convenience factor that drives patient satisfaction and health outcomes.[21]

Until recently, unless one was a part of an integrated delivery system, there was not a clear business case for patient portals. The combination of more feature-rich portals and relationship management systems with new reimbursement models that include patient satisfaction, as well as population health outcomes, make the business case very clear.

PREVENTING HOSPITAL READMISSIONS: THE CASE FOR MULTIPLE PERSONAL HEALTH IT TOOLS

The biggest savings from improved post-acute care is from the prevention of unnecessary hospital readmissions. This is a goal of the ACA. This is also key for financial success, as CMS instituted the Hospital Readmissions Reduction Program in October 2012 in which CMS will reduce reimbursement for providers with excess readmissions for acute myocardial infarction, heart failure, and pneumonia.[22]

Mastering transitions of care, from the inpatient hospitals to residential and sub-acute settings, is essential to preventing readmissions to hospitals.[23, 24, 25] Researchers at the University of Colorado Health Sciences Center developed a care transitions intervention that engages patients and their caregivers to be more active during care transitions, resulting in reduced rates of readmission.[26] The team found four key strategies that, together, bolster care transitions:

• Assistance with medication management;

• Timely follow-up with a patient's own primary and specialty doctors;

• Electronic medical records; and

• Patient education.

Personal health IT tools such as those discussed in this chapter address all of these areas. To prevent a hospital readmission for a patient discharged from the hospital with congestive heart failure, for example, a number of tools could be used together: a patient health portal to help the patient understand his/her condition; a medication adherence tool to track the complex medication dosing schedule; personalized emails between the patient or caregiver and the provider; a mobile health app to send text reminders; and a WiFi-enabled scale to make the provider aware of any dangerous spike in the patient's weight and adjust a medication dose, or schedule an immediate medical appointment.

PATIENT SATISFACTION: BEYOND OUTCOMES

Nearly all U.S. patients (90%) surveyed by Accenture wanted to use online self-service options such as accessing health information (83%), scheduling (72%), and prescription refills (72%).[27] Another survey of patients by Intuit found that 73% of U.S. patients would use a secure online tool that would make it easier to communicate with doctors' offices to ask care-related questions, schedule appointments, get lab results, and pay medical bills.[28] Such self-service options are expected by consumers, who seek to manage healthcare in the ways they manage other aspects of everyday life, from financial management to booking travel and movie tickets.

But health providers fall short of that retail experience Nirvana. In a nationwide survey of 6,000 consumers across nearly a dozen industries, provider customers (following health insurance customers) were the least likely to share a positive experience with others. Fifty-four percent of healthcare consumers talk about their positive experience, compared to 70% of retail and 66% of banking customers.[29]

In fact, Price Waterhouse Coopers (PwC) research discovered that consumers are less forgiving of health providers with whom they have had a negative experience, reporting that six out of ten negative experiences are more likely to be remembered longer in the health industry compared to other industry sectors.

In June 2004, testimony before the Subcommittee on Health of the House Committee on Ways and Means, Charles Safran, MD, a medical informatics expert, pointed to the role that patients can play in helping to address this problem, stating that, "Patients are the most under-utilized resource, and they have the most at stake. They want to be involved and they can be involved. Their participation will lead to better medical outcomes at lower costs with dramatically higher patient/customer satisfaction."

This is important for ACOs because customer loyalty can impact Medicare payment. As of October 2012, Medicare based reimbursements in part on patient satisfaction scores, cutting payments by 1%, or about $850 million for fiscal year 2013, under the Hospital Value-Based Purchasing program. Patient satisfaction scores will determine 30% of the incentive payments. "Leakage" of patients to a healthcare provider with no financial incentives to manage costs is a problem for an ACO, so the goal of bolstering customer satisfaction isn't just a "nice-to-do" objective: it's a critical success factor for ACO financial viability. Pioneer ACOs who shared lessons learned at panel discussion at the Health Datapalooza event in June 2012 highlighted the value of patient satisfaction and proactive communication in preventing leakage as a key to their ACO's success.

ACCOUNTABLE AND VALUE-BASED CARE REQUIRES COMMUNITY-BASED PATIENT ENGAGEMENT

Innovative providers, such as Eisenhower Medical Center, Geisinger Health System, and Joslin Diabetes Center are already responding to consumers' demands for more accessible, patient-centered care. These pioneering health systems are expanding their definitions of *where* healthcare is delivered: in the community, closer to where patients live, work, play, and pray, to use the phrase that Surgeon General Regina Benjamin, MD, has coined.[30]

It is not only consumers who seek greater health engagement: sponsors of health plans in both public and private sectors view engagement as a critical tactic to get patients more involved in self-care. Value-based benefit designs are increasingly adopted by employers providing health insurance to workers; these plans reward quality of care through payment incentives that are structured to discourage inappropriate, unnecessary, and costly care. Thus, the patient-facing health IT tools that are relevant for ACOs are also valuable in other value-based payment paradigms such as PCMHs, bundled payments, and payment-for-performance.

As care providers take on the mantle of accountable and value-based care, they must consider how to build community-based networks of care that leverage the precious asset of physician time to highest and best use. To do this will require team-based care, with specialists, nurses, allied health professionals, and a complement of community-based care providers. The critical success factor for effectively carrying out patient-centered, community-based accountable care will be an IT infrastructure—and the intelligence gained from that infrastructure—that brings patients into their key role as participants in their own healthcare.

CONCLUSION

Why do patient engagement tools to support ACO success matter for patients and consumers?

Patient engagement improves health literacy and health outcomes by getting people more involved in self-care.

- Patient engagement increases satisfaction with healthcare services and providers.

- ACOs that have strong health IT infrastructures and patient-facing tools can better track patient care and progress in managing chronic conditions, which can lead to fewer face-to-face visits to the doctor and more convenient and accessible care in the community.

Why do patient engagement tools to support ACO success matter for providers?

- Engagement leads to patients undertaking, as appropriate, more self-care at home, which can conserve the provider's time to see sicker patients. This means providers can function at their highest and best use.

- Well-designed patient engagement tools bolster health literacy, translating into more effective dialogue between providers and patients during exam time.
- When patient engagement improves health outcomes, providers at financial risk in ACOs can benefit from higher income and greater job satisfaction.
- Patient engagement tools can help "build" more activated patients, which can lead to greater job satisfaction for providers.

Why do patient engagement tools to support ACO success matter for the healthcare system as a whole?

- Patient engagement improves public health literacy, thereby improving a community's health profile.
- More activated and engaged patients have better health outcomes, which can lower costs of an episode of care.
- Having a "wired" health system in the community enhances the community's image as a modern health ecosystem.

For more information, read the following case studies at the end of this book:

- **Case Study 1:** Patient Engagement in a Patient-Centered Medical Home
- **Case Study 3:** How Social Media Has Changed My Medical Practice
- **Case Study 6:** Kaiser Permanente—My Health Manager
- **Case Study 8:** From God to Guide
- **Case Study 14:** Moving the Point of Care into Patients' Daily Lives via Text Reminders
- **Case Study 15:** New Models of Education for Pediatric Diabetes
- **Case Study 16:** Online Therapy and Patient Engagement
- **Case Study 22:** Veterans Health Administration, Department of Veterans Affairs—Patient Engagement and the Blue Button
- **Case Study 23:** Prescribing Apps for Weight Loss

REFERENCES

1. Millenson M. Building Patient-Centeredness in the Real World: The Engaged Patient and the Accountable Care Organization. Health Quality Advisors in collaboration with National Partnership for Women & Families, July 2012.

2. McClellan M, McKethan AN, Lewis JL, et al. A national strategy to put accountable care into practice. *Health Affairs.* May 2010; 29(5):982-990.

3. Anderson G. *Chronic Care: Making the Case for Ongoing Care.* Princeton, NJ: Robert Woods Johnson Foundation, 2010.

4. Centers for Disease Control and Prevention. Chronic Diseases: the Power to Prevent, the Call to Control. December 17, 2009.

5. Coleman EA, Parry C, Chalmers S, et al. The care transitions intervention. *Archives of Internal Medicine.* 2006; 25;166(17):1822-1828.

6. Charles Kennedy, MD, Head of Aetna's Accountable Care Services Division; interviewed by Jane Sarasohn-Kahn, MA, MHSA.

7. Finkelstein J, Knight A, Marinopoulos SP, et al. Enabling Patient-Centered Care through Health Information Technology. Evidence Report/Technology Assessment, Number 206. Prepared by The Johns Hopkins University Evidence-based Practice Center, under Contract No. 290-2007-10061-I. AHRQ Publication No. 12-E005-EF. Rockville MD: Agency for Healthcare Research and Quality, June 2012.

8. Manhattan Research. Taking the Pulse Physician Survey, 2010.

9. Deloitte Center for Health Solutions. 2008 Survey of Health Care Consumers.

10. Sarasohn-Kahn J. Primary Care, Everywhere: Connecting the Dots Across the Emerging Health Landscape. California HealthCare Foundation, November 2011.

11. 24/7 Primary Care. *Health Data Management*. March 1, 2010.

12. Zhou Y, Kanter M, Wang J, et al. Improved Quality at Kaiser Permanente Through Email Between Physicians and Patients. *Health Affairs*. 2010; 29:1370-1375.

13. Dixon R. Enhancing primary care through online communication. *Health Affairs*. 2010; 29:1364-1369.

14. Weinstock M, Hoppszallern S. Healthcare's Most Wired. *Hospitals & Health Networks*. July 2011; 85(7): 26-37.

15. Rukstalis M, Bloom F, Anderer T. Primary Care Web-Based Lifestyle Intervention for Type 2 Diabetes: Randomized Controlled Trial to Improve Knowledge and Self Care. Geisinger and dLife Presentation to 17th Annual HMO Research Network Conference, Boston, March 23-25, 2011.

16. Quest Diagnostics. Prescription Drug Misuse in America: Laboratory Insights into the New Drug Epidemic. Prescription Drug Monitoring Report, April 2012.

17. Versel N. Proteus gains de novo FDA clearance for ingestible biomedical sensor. *MobiHealth News*. July 30, 2012.

18. IMS Research. *Wireless Opportunities in Health and Wellness Monitoring – 2012 Edition*. June, 2012.

19. Sarasohn-Kahn J. The Connected Patient: Charting the Vital Signs of Remote Health Monitoring. California HealthCare Foundation, February 2011.

20. Conaboy C. 'Joslin everywhere' initiative to expand reach of hospital's diabetes care. *The Boston Globe*. February 28, 2012.

21. Osborn CY, Mayberry LS, Mulvaney SA. Patient Web Portals to Improve Diabetes Outcomes: A Systematic Review. *Current Diabetes Reports*. 2010; 10:422-435.

22. Kocher RP, Adashi EY. Hospital readmissions and the Affordable Care Act: Paying for coordinated quality care. *JAMA*. 2011; 306(16):1794-1795.

23. Coleman EA, Smith JD, Frank JC. Development and testing of a measure designed to assess the quality of care transitions. *Int J Integr Care*. 2002; 2(2).

24. Harrison A, Verhoef M. Understanding coordination of care from the consumers' perspective in a regional health system. *Health Services Research*. 2002; 37:1031-1054.

25. VomEigen KA, Walker JD, Edgman-Levitan S. Carepartner experiences with hospital care. *Med Care*. 1999; 37:33-38.

26. Coleman EA, Parry C, Chalmers S, Min S-J. The Care Transitions Intervention. *Arch Intern Med*. 2006; 166(17):1822-8.

27. Accenture. Is health care self-service online enough to satisfy patients? June 19, 2012.

28. Intuit. Intuit Health Survey: Americans Worried About Costs; Want Greater Access to Physicians. March 2, 2011.

29. PricewaterhouseCoopers. Customer experience in health care: The moment of truth. Health Research Institute. July 2012.

30. Brown W. Surgeon general discusses health and community. *Los Angeles Times*. March 13, 2011.

Chapter 3

How Meaningful Use Impacts Patient Engagement

By Jan Oldenburg, FHIMSS

INTRODUCTION

The closing days of 2008 were a dark time. With President Bush still in the White House and incoming President Obama waiting in the wings, the financial markets were melting down. The economy was headed into a deep recession that looked as if it could be a global depression. It was clear that action was needed—but what?

That historic confluence of difficult events generated an opportunity, as there was bipartisan commitment to create financial stimulus legislation to help pull the American economy out of the threat of depression. Legislators and staffers saw it as an opportunity to advance the long-term goal of getting more hospitals and physicians to implement electronic health record (EHRs). As early as the 1970s, when Senator Ted Kennedy asked the Office of Science in the Public Interest to investigate how information technology might impact healthcare, policy makers and practitioners alike had been dreaming of the day when healthcare would go digital. Like a mirage, however, the moment when the majority of providers would practice with an EHR always seemed to be five years away. The issue of health IT adoption, and its potential to positively influence healthcare quality and costs, was nonpartisan or bipartisan, with think tanks on both the right and left issuing white papers. Examples include former House Speaker Newt Gingrich who compiled best practices into a book entitled Paper Kills and its sequel, Paper Kills 2.0, with a foreword co-written by Speaker Gingrich, a Republican, and former U.S. Senate Majority Leader Tom Daschle, a Democrat.[1, 2, 3, 4]

Analyses at the time showed that the majority of EHR implementation value went to those who paid for healthcare, yet it was providers who were responsible for both paying for and implementing EHRs. That disconnect paralyzed forward progress on EHR implementation.[5, 6]

The economic stimulus package enabled concrete action to eliminate the paralysis. As the year wound down to a close and Inauguration Day for President Obama came closer, staffers on the Hill worked feverishly to develop a program to expedite the digitization of healthcare. The Health Information Technology for Economic and Clinical Health (HITECH) Act, enacted as part of the American Recovery and Reinvestment Act (ARRA) of 2009, was signed into law on February 17, 2009.

David Blumenthal, MD, then National Coordinator for Health IT, highlighted the goals of the HITECH Act and set the stage for meaningful use in a JAMA article in late 2009, when he said:

> "The provisions of the HITECH Act are best understood not as investments in technology per se but as efforts to improve the health of Americans and the performance of their healthcare system. The installation of EHRs is an important first step. But EHRs will accomplish little unless providers use them to their full potential; unless health data can flow freely, privately, and securely to the places where they are needed; and unless HIT becomes increasingly capable and easy to use."[7]

For the healthcare community, the stakes were enormous. The HITECH Act provided more than $27 billion in incentive payments for physicians and hospitals that implement EHRs and show that they are using them in meaningful ways. The payments are implemented over five years; at the end of five years, providers who have not implemented the capabilities face penalties of 1% of the Medicare physician fee schedule in 2015, up through a possible 5% in 2020. For hospitals, the payment structure is a bit different but no less significant. Each hospital can qualify for an initial amount, which is the sum of a $2 million base amount and the product of a per-discharge amount (of $200) and the number of discharges (for discharges between 1,150 and 23,000 discharges). After the first year, the payments are phased down annually.

CMS, which is responsible for making the incentive payments, takes direction and guidance from the National Coordinator, who in turn is advised by two Federal Advisory Committees (FACAs)—the HIT Policy Committee (HITPC) and the HIT Standards Committee (HITSC). These committees and their numerous workgroups are composed of stakeholders from across healthcare because Congress requires a balance of stakeholders to ensure "no single factor unduly influences" the recommendations of either committee.[8] There is a robust process for hearing and considering input regarding Meaningful Use requirements, in which the two FACAs make recommendations to the National Coordinator, who chooses whether to endorse the recommendations. CMS issues proposed rules, obtains public input, and then publishes a final rule for each stage of Meaningful Use. CMS produces one set of Meaningful Use requirements for Eligible Hospitals (EH), and another set for Eligible Professionals (EP). The requirements for each are further divided into a set of "core" (required) requirements and "menu" (a list from which a certain number must be chosen) requirements.

Indications are that this structure is working as designed. As of September 2012, more than 95,000 hospitals, physicians, and other eligible providers have received more than $4.1 billion in Meaningful Use incentive payments.[9] Hospital EHR adoption has doubled since the Meaningful Use rules were put in effect, going from 16.1% in 2009 to 34.8% by the end of 2012.[10] Current adoption on the part of EPs is at 55%, up from 48% in 2009.[11, 12]

PATIENT ENGAGEMENT

Meaningful Use has advanced key aspects of patient engagement. The concept of Meaningful Use was originally developed by the National Quality Forum (NQF), which released a report in 2008 identifying areas of healthcare that needed improvements.[13] Their ideas included improved population health, coordination of care, improved safety, increased efficiency, reduction of racial disparities, and patient engagement. These ideals were incorporated into five Health Outcome Priority Principles; every Meaningful Use objective is tied to at least one of these health outcome priority principles[14]:

1. Improve quality, safety, efficiency, and reduce health disparities.
2. Engage patients and families in their healthcare.
3. Improve care coordination.
4. Improve population and public health.
5. Ensure adequate privacy and security protections for personal health information.

While all of these priority principles work together for better population health, item number 2 directly impacts patient engagement. We will focus on the criteria around engaging patients and families in their healthcare for the remainder of this chapter.

This criterion was very important to the committee that created the Meaningful Use infrastructure. Paul Tang, MD, chair of the HIT Policy Committee's Meaningful Use Workgroup, Chief Innovation and Technology Officer and former Chief Medical Information Officer of the Palo Alto Medical Foundation, noted in an article about healthcare informatics that, "There is a clear value in direct input from patients and families. One of my favorite components of Meaningful Use is helping patients get access to their own data, and the next step may be developing tools to help them make use of it."[15] In an interview with Kate Christensen, MD, Dr. Tang further explained why this is his favorite part of Meaningful Use, saying:

• "First of all, it's their data anyway!

• It's also a spotting mechanism—enabling patients to view their own data helps to prevent things from falling through the cracks.

• Helps educate patients about their own conditions.

• We can provide "tools" that do more than give words of advice—they can encourage shared decision making and help patients to self-manage.

• Helps patients to take control and be more active in their own care."[16]

Charles Kennedy, MD, CEO of Aetna's Accountable Care Solutions division and another key participant in development of the Meaningful Use measures, echoed this sentiment when he said, "For deployment of core EMR technologies to achieve real value, they need to help achieve the triple aim of improving the experience of care, improving the health of populations, and reducing the per capita costs of healthcare. We felt it was important for patient engagement to be a key driver on the principle that chronic disease is a principal driver of cost of healthcare. Chronic illnesses are primarily managed in the patient's home rather than in clinical settings, and it isn't possible to achieve cost savings without engaging patients."[17]

One of the core goals of Meaningful Use is to make the patient (or the patient's advocate) the center of his or her own care team. The point of enabling messaging between patient and provider, allowing patients to access, view and transmit their clinical information is to enable patients to be the chief actors in their own healthcare. This goal plays out in the Meaningful Use stages as follows:

Stage 1: Build the base of electronic health records and establish limited capabilities—mostly in the optional or menu category—to provide patients with access to electronic versions of their records.

Stage 2: Expand support for patients' access to their health information by moving capabilities from menu to core, including the ability to securely message providers, and giving patients control over downloading and transmitting their records.

Stage 3: Focus on providing patient access to self-management tools, as well as providing patients and practitioners with access to comprehensive patient data.[18]

Many of the patient engagement capabilities required by Meaningful Use are already in use in progressive healthcare systems, but Meaningful Use, with its combination of incentives and penalties, is creating the expectation that providing access to personal health IT tools should be the norm rather than the exception in provider practices and hospitals.

STAGE 1

In Stage 1, Meaningful Use capabilities to support engaging patients and their families in their healthcare were mostly expressed as "menu" options, although there were three core objectives that supported patient engagement (more detail provided in Meaningful Use Chart 1 in Appendix B):

- Provide patients with an electronic copy of their health information on request.
- Provide patients with an electronic copy of their discharge instructions on request.
- Provide clinical summaries after each ambulatory visit.[19]

The patient engagement objectives in Stage 1 of Meaningful Use established a glide path toward providing useful personal health IT capabilities for patients. The Meaningful Use Committee used Stage 1 to signal that "providing patients with timely electronic access to their health information (including lab results, problem list, medication lists, medication allergies)"[20] is a critical goal of appropriate use of Electronic Medical Records.

Stage 1 began a significant transformation in the EHR landscape and the personal health landscape. Of the more than 110,000 eligible professionals (EPs) and over 2,400 eligible hospitals (EHs) that were paid over $3 billion in Medicare incentive payments by mid 2012, 39% or 11,215 of the EPs who attested chose the "timely electronic access to health information" menu option. If we assume that the average professional has a panel size of 2,000 patients,[21] that is 22,430,000 patients who had the ability to access their clinical data in 2012. Without Meaningful Use incentives, those 22.5 million patients might not yet have that capability.[22] Case Studies 2 and 13 describe the experience of small providers, leaders in implementing Meaningful Use, who have added patient engagement capabilities as a part of Meaningful Use Stage 1 attestation. Their experiences highlight how important it is for patient portal options to be included with or as inexpensive add-ons to EHRs; otherwise the expense is just too great. Larger providers, such as Kaiser Permanente, CHOP, and the VA (in Case Studies 6, 21, and 22) implemented patient engagement capabilities before Meaningful Use requirements came into play, but are stretching to be sure they address all of the requirements.

STAGE 2

Stage 2 Meaningful Use measures move patient engagement from the periphery to the core of Meaningful Use. They emphasize the requirement to provide patients with the ability to access, download, and transmit their data, and also highlight that to be successful, at least 5% of patients need to act on the availability of data. In addition, providers must have the capability for patients to communicate with them by secure messaging—and at least 5% of patients must actively engage in messaging.[23] See Meaningful Use Chart 2 in Appendix B for details of the Stage 2 requirements.

The framers of the requirements clearly understand that patient engagement does not exist in a vacuum. If providers are not engaged and motivated to help their patients understand their clinical data and engage with the health system, it will be harder for patients to be motivated and engaged. In this round of Meaningful Use, providers need both to implement the capabilities and also to market them and encourage their patients to use them in order to be successful.[24] Some commenters voiced concerns about making providers responsible for the actions of patients. In response, the patient-participation requirements were reduced from 10% to 5% of patients in the final rule, but they remain, signaling how important it is for providers to encourage patient participation.

The final Stage 2 regulations represent the process of change and growth in expansion of patient engagement capabilities. In order to meet these objectives, processes will need to change within hospitals and within clinics. There will need to be a new focus on workflow that is aimed at making valid data available to patients as quickly as possible. In addition, providers and systems will need to make patients aware of the capabilities that are at their fingertips and encourage their use. A provider who tells a patient in a visit, "I've ordered several lab tests for you, and if you sign up for access, you'll be able to get the results online much faster than I can get them out to you," or "I've prescribed a new medication for you; please email me in a couple of weeks and tell me how you're reacting to it," is much more likely to have patients who use electronic tools than a provider who is silent on the subject.

It is not enough, though, to just show patients their data. Data must be presented in the context of education and in ways that transform data into information—preferably actionable information. Meaningful Use Stage 2 requirements touch on this by requiring both professionals and hospitals to "Use clinically relevant information from Certified EHR Technology to identify patient-specific education resources and provide those resources to the patient if appropriate." This requirement—that data be presented to patients in terms that are understandable and with education about the data a click away—is essential to fulfilling the promise of the patient-facing Meaningful Use measures incorporated in Stage 2.

STAGE 3 (AND BEYOND)

Work is underway to form the Stage 3 Meaningful Use objectives and measures. As this book goes to print, the ONC's (Office of the National Coordinator) HIT Policy Committee has just released a request for comments regarding the Stage 3 definition of Meaningful Use of EHRs.[25] The full chart of current recommendations regarding engaging patients and their families is located in Appendix A, but highlights and commentary are provided below:

- Raising the percentage of patients who should receive preventive or follow-up care reminders from 10% to 20%;

- Requiring 30% of office visits to have electronic progress notes in the record (and considering a menu item that would enable patient access to those notes);

- Shortening the timeframe for EPs to make data available to patients from 4 days to 24 hours in most cases;

- Adding a menu item to enable 50% of patients to designate to whom and when a summary of care document is sent;

- Adding the ability for 10% of patients to submit patient-generated health information to improve performance on high-priority health conditions or improve engagement in care;

- Adding the ability to enable patients to request an amendment to their record online;

- Requiring the clinical summaries provided to patients to be pertinent to the office visit, not just an extract from the overall medical record;

- For the top five non-English languages spoken nationally, provide 80% of patient-specific education materials in at least one of those languages;

- Increase the percentage of patients who use secure messaging to communicate with EPs to 10%;

- Raise the requirement to record communication preferences to 20% of patients; and

- Add an EHR certification capability for connecting patients with clinical trials (use of this capability not required until future MU stages).

Sue Sutton, CEO and founder of Tower, a healthcare consulting group, notes:

Meaningful Use Stage 2 requires providers to interact with their patients digitally. While those interactions will soon be legislated, let's step back and map out a truly paperless, personalized patient experience from shopper to pre-treatment and treatment through post-treatment. Creating a true digital customer experience strategy will not only leverage those elements that are required by Meaningful Use, but those transactions will become part of overall personalized, disease-specific content that can be interactive and delivered using new media. We need to meet our patients in their comfort zone. They are used to self-service in other industry transactions. Why can't we provide a seamless, self-service interaction for patients that engages them while they are not inside the four walls of our facilities?[26]

If we had a magic wand to guide further Meaningful Use development, we would recommend some additional areas of emphasis for Stage 3 and beyond:

- It is positive that Stage 3 recommends that an electronic patient note be filed in 30% of office visits. The committee has asked for comments on whether they should include a menu item regarding exposing clinical notes to patients. **Recommendation:** After the positive success of the Open Notes project, we wholeheartedly endorse making this practice mainstream, and Meaningful Use Stage 3 would be an excellent way to make that a reality.

- We support the objective that would provide patients with the ability to designate to whom and when a summary of care document is sent. It is an important move to support patients' ability to obtain and provide a longitudinal view of their health record. It still, however, requires patients to act as their own Health Information Exchange (HIE), responsible for gathering all of their records from providers and eliminating duplications themselves. While Stages 1 and 2 have made the data digital and mandate the ability to transmit the data, Stage 3 would do well to incorporate mechanisms for the simple, transparent aggregation of patient data. Accountable Care Organizations (ACOs) and Patient Centered Medical Homes (PCMH) will support integration of care, but today many patients still participate in a fragmented system where pieces of their records are stored at the offices and in the systems of many different providers, making it very complex to assemble a complete patient record. **Recommendation:** Action is required to make HIE far easier for both patients and providers. We recommend the Policy Committee enable the formation of comprehensive, longitudinal community patient records for both patient and provider access.

- ONC's recent support of consumer-mediated exchange may encourage development of this supporting infrastructure. Jim Hansen, of Lumeris (formerly of the Dossia Consortium), who was an active participant on the Health Information Technology Standards Panel (HITSP) and workgroup leader on the ONC HIT Standards Committee Consumer Power Team, had this to say in an email exchange about the role of the Direct protocol in facilitating consumer information exchange:

"Direct will play two essential roles in dramatically increasing the 'information liquidity' that is needed to transform our health and healthcare system. First, Direct allows small and medium sized practices to participate at a dramatically lower cost and complexity level than alternative HIE approaches. Second, Direct allows consumer/patients, if they so choose, to finally have the tools to act in the role of what Dr. Brad Perkins [MD] (Dossia Chairman, Vanguard Chief Transformation Officer) calls an 'HIE of one' to collect, aggregate, utilize and share their health information without depending on other's desire/decision to participate in other forms of HIE. Principally for these two reasons, it is truly a game changer. Many of my colleagues expect that the next round of HIE standards will incorporate both Direct ('push approach') and NwHIN ('push & pull approach'), allowing a seamless migration and interconnectivity between the two."

- We support the likely Stage 3 initiative to encourage providers to incorporate and use patient-collected data in clinical care, but we would like to see it accelerated. Patient-collected data are likely—soon—to be gathered from devices as an outgrowth of ambient-sensing device development and use. Quality care for patients and the practice of medicine will be advanced when physicians can incorporate information from a real-time stream of biometric data and data about activities of daily living. The ability to incorporate, aggregate, and analyze such data—along with data from external factors such as temperature, medication levels, allergens, and stress levels, may bring the practice of medicine to a completely new level. Market activities are currently exploring this and testing practices for incorporating or summarizing this resulting "big data" for clinical and patient use. **Recommendation:** Strengthen the proposed Stage 3 recommendation to incorporate patient-provided data by adding the ability to import data from remote sensing/monitoring devices. **Recommendation:** Move from the retrospective analysis of data to the ability for both providers and patients to take action "in the moment" based on data. This is not currently incorporated in the proposed objectives for Stage 3.

- We also fear that the Meaningful Use focus on incorporating patient-centered functionality within the EHR may give rise to monolithic systems that do everything—but none of it particularly well. We realize that the framers may have intended to move healthcare organizations away from "best of breed" technologies where millions of dollars are spent on interfaces rather than on implementation. In the realm of patient engagement, where attention to design and to the user interface is critical to widespread adoption, this might end up being a limiting rather than enabling factor. **Recommendation:** Make it easier for organizations to get Meaningful Use certification for sharing data and communicating with patients using third-party tools that really focus on usability and enhanced patient experience—not an area of core competence for most EHR vendors.

An Interview with Dr. Charles Kennedy

Charles Kennedy, MD, is CEO of Aetna's Accountable Care Solutions division and one of the members of the HIT Policy Committee responsible for creating the Meaningful Use requirements and making patient engagement a key component of them. When asked to reflect on how patient engagement became a key element, Dr. Kennedy was thoughtful. He noted that the HIT Policy Committee wanted to have a significant impact on healthcare in the United States. At the same time, the capabilities they promoted needed to have significant economic impact, as they were being created as part of ARRA with the goal of spurring employment and business investment. The two objectives did not completely align, because the EHR capabilities available at that time did not have a proven track record in improving population health and quality or reducing cost of healthcare. The HIT Policy Committee also did not want to simply create incentives for purchasing technology, though that might have achieved the economic aims of the legislation. To have the impact they wanted on healthcare cost and quality, the tools had to be used.

That is how the framework of Meaningful Use was born: create requirements and a certification process to ensure that the technology on the market would be mature enough to impact healthcare cost and quality, then build a structure to ensure that professionals and hospitals were actually using the technology in ways that it would have the desired impact. Stage 1 of Meaningful Use was really about ensuring a solid base of capabilities.

As they moved into Meaningful Use Stage 2, Dr. Kennedy noted, the HIT Policy Committee focused on turning the core technologies into real value that could begin to achieve the triple aim: improving the experience of care, improving the health of populations, and reducing per capita costs of healthcare. Several members of the Committee pushed for patient engagement to be one of the next drivers on the principle that a primary driver of cost of healthcare is chronic disease. Chronic illnesses are primarily managed in the patients' homes rather than in clinical settings, and it is not possible to achieve cost savings without engaging patients and their families. Another key driver was care coordination—how to get physicians and other members of the care team to coordinate care effectively in order to reduce duplicate tests and procedures and improve patient transitions in care, all components that drive up healthcare costs when done poorly. The final driver was quality. Once providers have the primary EHR platform, they must be able to monitor and measure the quality of their outcomes.

When Dr. Kennedy considers using healthcare data to create value, whether for doctors or patients, he has some consistent measures: It has to be useful, timely, accurate, contextually relevant, and actionable. Today we cannot use existing data for useful, timely, accurate, contextually relevant, and actionable treatment activities because we either do not have digital data or because what we have is fragmented.

A patient portal deployed in a small primary care doctor's office can provide useful data, but it is an incomplete view of most patients' medical care. To really provide value, all the data about the patient need to be put into context for both the patient and the physician. The patient's data also need to be surrounded by education to ensure that it is both understandable and actionable.

Outside of integrated delivery systems, the system constraints that create siloed and fragmented views of a patient's medical context cannot easily be overcome, and this dilemma drove extensive conversations in the Committee. The Committee's goal is to create a network effect so that every provider and every patient's data becomes a part of the whole—building, over time, a virtually integrated system across the data that enables patients and providers alike to view health data that is useful, timely, accurate, contextually relevant, and actionable.

Dr. Kennedy believes that the future of healthcare lies in collecting data from disparate locations, bringing it into a central repository, and deriving meaning from it with clinical ontology, semantic interoperability and advanced analytics. Dr. Kennedy is hoping that by Stage 3 and 4 of Meaningful Use, the focus will be on these advanced capabilities that can accelerate systems change.[27]

CONCLUSION

Meaningful use has done a great deal to drive the adoption of electronic health records and provide consumers with access to their data in a meaningful way. Although there is still a significant distance to travel, Meaningful Use is helping to transform healthcare for patients, providers, and the system as a whole.

Why does Meaningful Use matter for patients and consumers?

- It establishes incentives for providers to give patients access to their clinical information.
- It establishes a requirement for providers to set up secure messaging capabilities with patients, establishing a 21st century, provider-patient communication channel.
- It requires providers to supply educational materials that are specific to the patient's situation.

Why does Meaningful Use matter for providers?

- It establishes standards for implementation of electronic health records and defines what it means to use them in meaningful ways.
- It establishes a framework of incentives for providers who implement the standards before the end of the program and penalties for those who do not.
- It integrates information technology into the practice of medicine.

Why does Meaningful Use matter for the healthcare system as a whole?

- It moves the healthcare industry forward on dimensions of cost and quality.
- It enforces standards for adoption and use of technology in medicine.
- It provides patients with access to their own data.

For more information, read the following case studies at the end of this book:
- **Case Study 2:** Innovation in the Context of Meaningful Use
- **Case Study 13:** Meaningful Use Implementation and Patient Activation

REFERENCES

1. Kendall DB. Protecting privacy through health record trusts. *Health Affairs*. 2009; 28:2:444-446.

2. Haislmaier EF. Health care information technology: Getting the policy right. Posted June 2006. http://www.heritage.org/Research/HealthCare/wm1131.cfm. Accessed September 28, 2012.

3. Merritt D. P*aper Kills: Transforming Health and Healthcare with Information Technology.* Marietta, GA: Center for Health Transformation; 2007.

4. Merritt D. Paper Kills 2.0: *How Health IT Can Help Save Your Life and Your Money.* Marietta, GA: Center for Health Transformation; 2010.

5. Miller R, West C, Brown T, et al. The Value of Electronic Health Records in Solo or Small Group Practices. *Health Affairs*. 2005; 24,5:1127-1137.

6. Pan E, Johnston D, et al. *The Value of Healthcare Information Exchange and Interoperability.* Boston: Center for Information Technology Leadership; 2004.

7. Blumenthal D. Launching HITECH. New Eng J Med. 2010; 362:382-385, Posted February 4, 2010. http://www.nejm.org/doi/full/10.1056/NEJMp0912825. Accessed September 23, 2012.

8. American Recovery and Reinvestment Act of 2009: Title XIII – Health Information Technology. HR1. 111th Congress. 2009.

9. Holland, Elizabeth. Medicare and Medicaid EHR Incentive Programs Report to HIT Policy Committee. www.healthit.gov/sites/default/files/holland_hitpc_october-2012.pdf. Posted October 3, 2012. Accessed November 2, 2012.

10. Charles D, Furukawa M Hufstader M. ONC Data Brief. Posted February, 2012. http://www.healthit.gov/media/pdf/ONC_Data_Brief_AHA_2011.pdf. Accessed September 23, 2012.

11. Jamoom E, Beatty P, Bercovitz, A, et al. Posted July 2012. http://www.cdc.gov/nchs/data/databriefs/db98.htm. Accessed September 23, 2012.

12. Gur-Arie M. EHR Adoption Rates. Posted December 1, 2012. http://onhealthtech.blogspot.com/2011/12/2011-ehr-adoption-rates.html. Accessed September 23, 2012.

13. National Priorities Partnership. http://www.qualityforum.org/Setting_Priorities/NPP/National_Priorities_Partnership.aspx. Accessed September 23, 2012.

14. Department of Health Care Services. EHR Final Rule. www.dhcs.ca.gov/Documents/OHIT/EHR%20FINAL%20RULE.pdf. Accessed September 23, 2012.

15. Raths D. Meaningful Use: First Steps. *Healthcare Informatics*. August 29, 2010. Posted August 27, 2010. www.health care-informatics.com/article/meaningful-use-first-steps. Accessed September 23, 2012.

16. Paul Tang, MD; interviewed by Kate Christensen, MD, July 11, 2012.

17. Charles Kennedy, MD; interviewed by Jan Oldenburg, July 30, 2012.

18. Mostashari F, Tavenner M. Stage 2 Meaningful Use NPRM Moves Toward Patient-Centered Care Through Wider Use of EHRs. Posted Date February 2012. http://www.healthit.gov/buzz-blog/from-the-onc-desk/stage-2-meaningful-nprm. Accessed September 23, 2012.

19. *Federal Register*. Medicare and Medicaid Programs; Electronic Health Record Incentive Program. July 28, 2010. Pages 44313 - 44588 (276 pages). [FR DOC #: 2010-17207].

20. Ibid.

21. The doctor will see you if you're quick. Posted April 16, 2012. http://www.thedailybeast.com/newsweek/2012/04/15/why-your-doctor-has-no-time-to-see-you.html. Accessed September 23, 2012.

22. CMS Electronic Health Record (EHR) 2011 Incentive Program Eligible Professionals Public Use File (PUF) Data Dictionary and Codebook. Published February 2012. *ebookbrowse.com/ep-puf-datadic-cb-pdf-d344729117*. Accessed September 23, 2012.

23. *Federal Register*. Medicare and Medicaid Programs; Electronic Health Record Incentive Program Stage 2, September 4, 2012. 77 FR 53967, 53967 -54162 (196 pages).

24. *Federal Register*. Medicare and Medicaid Programs; Electronic Health Record Incentive Program Stage 2, September 4, 2012. 77 FR 53967, 53967 -54162 (196 pages).

25. Office of National Coordinator of Health Care IT, Request for Comments on the Stage 3 definition of Meaningful Use of EHRs. Posted November 16, 2012. http://www.healthit.gov/sites/default/files/hitpc_stage3_rfc_final.pdf. Accessed November 21, 2012.

26. Sue Sutton, CEO and founder, Tower; interviewed by Jan Oldenburg, September 15, 2012.

27. Charles Kennedy, MD; interviewed by Jan Oldenburg, July 30, 2012.

Chapter 4

Patient Safety and Risk Management: Better Information for Physicians Means Reduced Risks (and Better Outcomes)

By Kate T. Christensen, MD

"To me, one of the principles of safe care is that it's a team sport, and one of the most neglected members of the team is the patient."

–Matt Handley, MD,
Medical Director for Quality and Informatics, Group Health Cooperative

INTRODUCTION

Electronic health records (EHRs) play a critical role in patient safety in hospitals and clinics.[1] Less familiar is the role personal health information technology tools linked to a physician's practice play in enhancing risk management and patient safety in medical care. In its 2001 report, "Crossing the Quality Chasm," the Institute of Medicine included the following recommendation for improving quality care: "Patients should have unfettered access to their own medical information and to clinical knowledge. Clinicians and patients should communicate effectively and share information."[2] This chapter will explain how this statement applies to patient safety.

There are several ways that patient engagement using personal health IT can provide better information for physicians and their care teams in their daily practice.

- It starts with the benefit of having a new patient bring or send an electronic version of their medical history. It could be in the form of a Continuity of Care Document (CCD) with the medical history, an ASCII file from the Blue Button initiative, or a PDF file. The same type of document can provide for physicians a deeper understanding of treatments received from other physicians or as the result of out-of-area care.

- When patients have online access to their medical records, they can also provide sometimes-critical updates or corrections. By partnering with patients to create more complete and error-free medical records, physicians can reduce the risk of medical errors and malpractice claims.

- Providing automated access to lab test results online additionally reduces the risk that abnormal results go unnoticed or are not communicated to patients.

- Finally, patients who track measures like blood sugars, blood pressures, or peak flows between visits with biometric devices can provide for physicians a much better understanding of their health status than the occasional office-based measurement.

PATIENT ACCESS TO PORTABLE FORMS OF THE HEALTH RECORD

A physician sees a new patient, an elderly man with a history of migraines, who is now having a terrible time with frequent episodes. He is taking only an over-the-counter migraine medication. He does not remember any of his prior prescription migraine medications, but says he has been on "all of them." He does not remember why he was taken off them, but he recalls that one gave him chest pain. It would be a challenge to treat this patient without having his medication history, and relying on guesswork could lead to harm.

As new patients come into a medical practice, they often are not able to bring their healthcare records with them. This can mean starting from scratch. Because memories fade, and important details may get lost with time, errors can be created in the new record that could lead to patient harm down the road.

A solution to this problem is to provide for patients a portable version of their personal health records, so they can give these records to any provider who needs it. The availability of a summary record in a portable format has been given a big boost from Meaningful Use standards. Stage 1 Meaningful Use rules required eligible professionals to provide patients with an electronic copy of their data upon request. Stage 2 rules require that the system allow patients to download or transmit a copy of their health record. The Stage 2 requirement that this capability be available as a self-service feature will surely increase adoption. One large-scale example can be found at the Veterans Administration's (VA) healthcare system. The VA allows registered patients to download a summary of their health record from the patient portal, using their "Blue Button" feature. At Kaiser Permanente, members can create a PDF document from the patient portal, which they can send via email to a new physician.

The portable record still has the risk of being a static document; it is only a snapshot of one point in time in the record and can still be out of date. New patients also could bring in printouts from their previous provider, but it can be harder to find relevant data in a pile of papers. Giving patients access to a portable version of their own record can significantly help in providing a complete medical history and in preventing avoidable medical errors.

PATIENT UPDATES AND CORRECTIONS TO RECORD

There is a common concern among physicians that showing a patient his/her chart can create unwanted problems. Errors in the medical record can include the wrong patient's data, errors in history-taking, misunderstandings between patient and doctor, and outdated medication lists. According to some physicians, asking for feedback on the medical record could open a Pandora's box of complaints, requests to custom-edit the chart for a variety of reasons, and increased litigation.

Fortunately, there are a growing number of physician practices that provide online access to at least part of the medical record for adult patients. They also provide ways for patients to provide notices to the practices about updates, corrections, and care received elsewhere. We can learn from their experience.

Patients can submit requests for corrections by email when available, in writing, and, in some practices, through the patient portal. Requests can be vetted by office staff or the physician to ensure that the chart remains accurate.

Paul Tang, MD, is the Chief Innovation and Technology Officer and former Chief Medical Information Officer of the Palo Alto Medical Foundation. In an interview with the author, Dr. Tang explained that, "By providing patients with access to their own records online, patients contribute to their care and safety by helping to maintain the completeness and accuracy of their information."[3]

Medication list errors can be especially risky, and given the frequency of medication changes for many patients, it can be especially important for patients to be able to provide updates to their medication list. According to a 2006 report by the Institute of Medicine, at least 1.5 million people are harmed each year by medication errors.[4] Maintaining an accurate medication list will help to lessen this burden.

The potential patient safety benefits to the physician practice are yet to be realized on a widespread basis. By showing patients their medical records and providing ways for them to provide corrections or updates, to be verified by the clinical team, the record can become more accurate and medical care will be safer. One challenge to be overcome is the unstructured nature of the data provided from the patient. Until further enhancements are made, the data may not automatically populate the right places in the chart, so some effort is involved in integrating the new data with the existing chart record. However, this is a problem that the appropriate use of technology can overcome.

PATIENTS PROVIDING A 'SECOND SET OF EYES' FOR ABNORMAL TEST RESULTS

A 2009 study published in the *Archives of Internal Medicine* found that approximately 7% of clinically significant abnormal test results were not communicated to patients in three healthcare systems using EHRs.[4] What they did not look at, however, was the potential to reduce that percentage by giving patients access to their own results, thus providing a safety net.[5,6,7] As Dr. Tang explained, "One doctor keeping track of a whole panel of patients is not as reliable as each patient reviewing their own data. This spotting mechanism helps to prevent things from falling through the cracks. Transparency is a path to safer practice that at the same time delights patients."[8]

Matt Handley, MD, Medical Director for Quality and Informatics, Group Health Cooperative, noted why this is important in an interview with the author. "To me, one of the principles of safe care is that it's a team sport, and one of the most neglected members of the team is the patient. Having patient-centered health IT allows you to bring the patient into the team in the way you could not do otherwise. The more they see, the more they understand, and can double-check. That includes all kinds of results to make sure they've been attended to. It also helps makes sure they understand the plan.[9]

I know of cases in which the patient identified an important finding online which had slipped through the cracks. Examples include positive cancer screening tests such as mammograms or colorectal cancer screens that went unheeded by the team. We would have eventually caught it with a safety net system, but the patient caught it right away or shortly after the incident and that jump-started the work-up. We have also seen elevated creatinine with pending renal failure caught by a patient that allowed adjustment of medications and evaluation that was clinically important."[10]

Viewing Radiology Results Online

"We know that there will be a time when a result goes unmanaged," said Matt Handley, "even if it's seen by a physician, especially if there's something in the body of a report that isn't mentioned in the conclusion. A patient may say 'But what about that nodule they saw in my lung?' and point out something in the detail of a report. We are confident that the patient will find some things in a report that a busy clinician might occasionally miss if they are not called out by the radiologist. We think it's important to have as many people as possible reading things carefully to make sure nothing slips through the cracks.

"We also already know that there are fewer patient phone calls when you release radiology results than when you don't. We got a lot of calls from people wondering about their imaging results, and some of those calls go away when they can see them online. We had the same experience with lab tests; we had an awful lot of people calling to find out about their lab tests, but those calls generally went away when they could access their test results themselves.

"Our primary care docs say that someone asks on average once or twice a month about the medical terminology in an imaging result, but it's not frequent. We've found the same holds true for lab test results."[10]

Project Open Notes

Recently, collaborators from Beth Israel Deaconness Medical Center, Geisinger Health System primary care clinics, and the Harborview Medical Center sought to improve patient care by inviting patients to read their doctors' notes.[11] The centers sampled patients receiving care in different settings: an urban academic setting, rural integrated health settings, and a county hospital respectively. Over 90,000 patients under the care of 114 primary care physicians (PCPs) were invited to review encounter notes during a year-long demonstration project. Notes were only available to the patient after the doctor had signed the note. The electronic signature triggered secure messages to patients, letting them know they could logon to the facility portal and view the note. Jan Walker, the study lead says, "On one hand, this seems groundbreaking and on the other hand, patients have a legal right to their records; so what we did was simply make it easier for patients to access those notes."

The research team was interested in evaluating the perceptions of both doctors and patients toward the project, which was labeled "Open Notes," and found that their opinions varied widely. The elegant study design allowed researchers to collect pre-test survey data from both doctors and patients who elected to and those who declined to participate in the project. Patients were overwhelmingly positive about the idea of reading their encounter notes and were much less concerned (fewer than one in six) about being worried or confused about reading the note. Their results suggest that all patients, young and old, educated and uneducated, good health and poor health, anticipate that reading the doctors' note may help them. Doctors were far more worried about the downside of the project, citing concerns about the extra time it would take to create the notes and address patient questions outside of visits, and the lack of freedom to be candid in writing the notes. A large percentage (36% to 62%) of participating physicians and non-participating physicians (18% to 33%), however, thought that patient care would be safer if patients could see their notes and that patient satisfaction would also improve.

A follow-up article, published in September 2012, describes the post-test survey results. Of the patients surveyed, "77% to 87% across the 3 sites reported that open notes helped them feel more in control of their care; 60% to 78% of those taking medications reported increased medication adherence." The paper concludes by saying, "Patients accessed visit notes frequently, a large majority reported clinically relevant benefits and minimal concerns, and virtually all patients wanted the practice to continue." Ms. Walker added, "At the end of the year, we went back to participating PCPs and told them we were done, so if they wanted to stop publishing their notes, they could. Not one single doctor dropped out."[12]

Anecdotal stories highlight the important patient safety aspects of this project. One reluctant doctor turned into a big fan, commenting that he "felt safer" knowing the patient could read his office notes and double-check that things were accurate. Patients comment that they simply cannot remember everything that happens in the exam room, so being able to look up information online after the visit is a huge gain. Ms. Walker tells us that every time she talks about Open Notes, someone inevitably comments about how great it would be to look up an encounter note after an elderly parent in another city is seen by their doctor.

One patient thought he remembered his doctor telling him that three tests would be ordered but the patient only received appointments for two. The patient looked up the note online and called the office a week later, asking "Did you forget to order that third test?" The office had failed to make the third appointment but did so promptly following the call.

Another patient was distracted with an issue related to her pregnancy when she saw her PCP and forgot to follow up with a specialist as instructed. A nagging feeling led her, two weeks later, to go online and read her note. The PCP had wanted her to see a dermatologist about a skin lesion. She made the appointment and learned that the lesion was dangerous and had to be removed.

Several patients made comments regarding the impact of seeing something written down in "black and white" and how that makes the issue "real" to them. Ms. Walker said, "Suddenly patients 'get religion' about what they are supposed to do."

The team hopes to expand the concept of patient access, giving patients an easy way to view more and more of their own health records. The culture change that is required to make this a reality, it seems, has already happened.

Sigall Bell, MD, one of the investigators on Open Notes, added that shared visit notes can improve patient safety through more accurate histories, medication lists, and potentially even attenuated diagnostic delays. "Study patients found mistakes in their notes, including documentation to increase the dose of a medication the patient did not take. The patient was able to help her physician correct this mistake." She also gives the following example about diagnostic delays: a patient in the study reported that with Open Notes her breast cancer could have been diagnosed earlier, as a previous doctor had documented "the exact area" and a concern about breast cancer, but did not tell the patient about any concern.[13]

The Importance of Sharing Care Plans

A Kaiser Permanente patient stated, "Having my medical records (lab results) available to me online gives me the opportunity to review, research and develop thoughtful questions for my doctor. Our following discussion then is referencing the same data and we have the opportunity for a two-way, productive dialogue. The door is then open for me to play an active role in any necessary follow-up actions."[14]

Dr. Handley agreed: "One of the biggest errors in medicine is when the care team has a very explicit medical plan that they believe they have communicated effectively to the patient and then they later find that the patient did not understand the plan. Having the care plan shared transparently is a huge advantage to the clinician and the patient. Most people lose the piece of paper with the summary, so the ability to access past visit summaries at any time from any place is an important benefit to the patient. It makes sure that the care plan is truly theirs.

"I also have had patients who I thought we were on the same page about something, but later, after viewing the summary online, recognized that they didn't understand something they thought was clear in the exam room. This includes important questions about medication strategies, auto-titration, and managing insulin. I thought I had explained it, they thought they understood it, but when they read about it later they realized they were missing important information."[15]

Patient-Generated Health Data Provide Alerts If a Critical Metric Is Recorded

"Patient safety is part of a quality continuum," said Dr. Tang. "Providing personal health IT to patients not only helps us ensure the accuracy of the information in their chart, such as the medication list or family history, but also facilitates collection of new information, such as patient-entered data like functional status or glucometer readings. Having these data provides new information that can help us personalize treatment recommendations to individual patients. Skeptics might argue that this takes too much time, but often we learn information about a patient that can prevent a visit or help us intervene before a health condition gets worse. By triaging the incoming information, some of the updates, such as a prescription renewal request or flu shot update, can be handled by other members of the professional health team."[16]

There are many potential avenues and types of data that patients are able to generate, or will be able to generate in the near future. Many people use devices now to track their fitness activities and/or caloric intake that can be uploaded or manually input into their personal health record or another tracker. As described in a recent issue of *The Economist*, some go far beyond that and quantify multiple facets of their daily life on a second-by-second basis. "They are an eclectic mix of early adopters, fitness freaks, technology evangelists, personal-development junkies, hackers and patients suffering from a wide variety of health problems. What they share is a belief that gathering and analyzing data about their everyday activities can help them improve their lives—an approach known as "self-tracking," "body hacking" or "self-quantifying."[17] That data are largely for their own benefit, and presumably does not go into their medical records. It is easy to imagine a future in which wearable monitors for people with chronic diseases will allow data to flow automatically to their doctors; data analysis tools would analyze the data and notify a caregiver in real time if the patient is experiencing an issue that requires intervention.

When it comes to condition-related biometrics, such as blood pressure, blood sugar, and weight, the context will dictate whether or not the data need to be seen by the physician or care manager and be recorded in the chart. For example, data from someone with mild, controlled hypertension who tracks her blood pressure twice a day would not usually need to be reviewed by a professional. But if that person had labile hypertension, was pregnant, was on a new medication, or needed close monitoring for any other reason, then it would make sense to have the data flow to the chart and a healthcare professional for review and action if needed. This is another area that requires both efficient office workflows and alerting and sorting software to make sure that critical values are addressed emergently.

Online questionnaires are another very useful format for gathering patient input in an organized way, and they can be utilized for a variety of purposes. For example, a questionnaire sent out by email before a visit can gather "history of present illness" information in a way that is more efficient for the office and the physician. It allows them to screen for patients who need to be seen urgently and spend less time in the exam room gathering basic information. Follow-up questionnaires after treatment or a visit can reveal problems in care or misunderstandings about the care plan, which can lead to safer care.

When patients contribute critical health data, their providers receive a more complete picture of their health and conditions, which can lead to better care decisions. For example, if they are aware that a patient's blood pressure is often elevated after 6 pm, physicians can adjust the patient's medications accordingly.

What is the downside? Citing a common fear that providers might become overwhelmed with data, Doug Bonacum of Kaiser Permanente said: "Engineers must make technology smarter. Yes, there is such a thing as too much data. Computer algorithms must be built to alert physicians to alarming trends in data (e.g., weight increasing in heart failure patients, labile blood pressure values in the hypertensive, glucose readings that reach critical limits in diabetics), instead of requiring physicians to sift through millions of data points. Alerts would prompt us to action when a rescue was required."[18]

CONCLUSION

Future progress will likely include widespread access to the entire record, inpatient and outpatient, including progress notes, and more easily accessible and transportable records via mobile devices.

Why do patient safety improvements using personal health IT (PHIT) matter to patients?

- Patients benefit from safer care when their physicians have an accurate record of their medications, conditions, and allergies.
- When the record is shared, patients are more likely to catch abnormal test results that require follow-up, thus avoiding the harms of delayed diagnosis and treatment.

Why do patient safety improvements using PHIT matter to providers?

- Increased patient access to their own health data improves patient safety for any kind of practice. It does so by creating a more complete record for a patient new to your practice, improving the accuracy of the chart, creating more complete health data in the chart with patient-generated data, and engaging patients more actively in their care.
- Providers reduce their risk of preventable medical errors when they share clinical data with their patients.

Why do patient safety improvements using PHIT matter to the healthcare system as a whole?

- Widespread implementation of PHIT holds the promise of reducing medical harm caused by inaccurate records and missed abnormal test results. Avoidable harm is very costly, in terms of personal suffering, as well as the financial burden for our society.

For more information, read the following case studies at the end of this book:

Case Study 4: From the Heart: The Intersection between Personal Health and Technology

Case Study 7: Everyone Performs Better When They Are Informed Better

Case Study 9: Restoring Trust with Information

Case Study 11: The Power of Information and the Partnership of Trust

REFERENCES

1. Boancum D. Improving Patient Safety. In Liang L (ed). *Connected for Health: Using Electronic Health Records to Transform Care Delivery*. San Francisco: Jossey-Bass; 2010: pp 157-178.

2. Institute of Medicine. Crossing the Quality Chasm: A New Health System for the 21st Century. Washington, DC: National Academy Press; 2004.

3. Paul Tan, MD; interviewed by Kate Christensen, MD, July 11, 2012.

4. Committee on Identifying and Preventing Medication Errors. Front Matter. Preventing Medication Errors: Quality Chasm Series. Washington, DC: The National Academies Press; 2007.

5. Casalino LP, Dunham D, Chin MH, et al. Frequency of failure to inform patients of clinically significant outpatient test results. *Arch Intern Med.* 2009;169(12) 1123-1129.

6. Christensen K, Oldenburg J. Giving patients their results online might be the answer. *Arch Intern Med.* 2009; 169(19):1816-1817.

7. Davis GT, Singh H. Should patients get direct access to their laboratory test results?: An answer with many questions. *JAMA.* 2011; 306(22):2502-2503.

8. Paul Tan, MD; interviewed by Kate Christensen, MD, July 11, 2012.

9. Matthew Handley, MD; interviewed by Kate Christensen, MD, July 16, 2012.

10. Matthew Handley, MD; interviewed by Kate Christensen, MD, July 16, 2012.

11. Walker J, Leveille S, Ngo L, et al. Inviting patients to read their doctors' notes: Patients and doctors look ahead. *Ann Intern Med.* 2011; 155(12): 811-819.

12. Tom Delbanco, Jan Walker, Sigall K. Bell, et al. Inviting patients to read their doctors' notes: A quasi-experimental study and a look ahead. *Ann Intern Med.* 2012; 157(7):461-470.

13. Sigall Bell, MD; interviewed by Kate Christensen, MD, July 31, 2012.

14. Matthew Handley, MD; interviewed by Kate Christensen, MD, July 16, 2012.

15. Diane McNally; interviewed by Kate Christensen, MD, July 25, 2012.

16. Paul Tan, MD; interviewed by Kate Christensen, MD, July 11, 2012.

17. *The Economist.* The quantified self: Counting every moment. March 3, 2012.

18. Douglas Bonacum; interviewed by Kate Christensen, July 11, 2012.

Chapter 5

Engaging Patients with Personal Health IT for Quality

By Jonathan Wald, MD, MPH, FACMI

INTRODUCTION

Quality has been a primary concern in healthcare since the Institute of Medicine's 2000 (IOM) report, "To Err Is Human" identified preventable medical errors as a major cause of death and health risk for patients.[1] Taking a systems view toward understanding and engineering quality into healthcare, the IOM's 2001 "Crossing the Quality Chasm"[2] report identified 6 aims and 13 recommendations that should guide improvement efforts. The focus on quality in medical care intensifies whenever efforts to constrain medical practice increase,[3] such as through cost containment efforts, public health programs, practice variation reduction, and practice redesign—including technology initiatives. Substantial health reform efforts underway through the American Recovery and Reinvestment Act (ARRA) and Patient Protection and Affordable Care Act (ACA) are increasing the focus on cost containment, quality, and the patient experience.

Despite a sustained focus on measuring and improving quality over more than a decade, it is difficult to define precisely. One widely-accepted definition of quality from the IOM's "Crossing the Quality Chasm" report is "The degree to which health services for individuals and populations increase the likelihood of desired health outcomes and are consistent with current professional knowledge."[2] Quality is complex for several reasons. First, different stakeholders define it differently. From a patient perspective, a trusting relationship, a delightful office visit experience, a reduction in procedural risk, an accurate diagnosis, and an appropriate and safe treatment choice are all indicators of quality. To a provider, reducing no-shows, adhering to clinical guidelines, thorough documentation, and improvement in Healthcare Effectiveness Data and Information Set (HEDIS) measures are all measures of quality and illustrate the broad range of quality indicators. Payers and public policy leaders measure quality in terms of reduced hospitalization rates and increased patient activity levels. Quality is both what the healthcare system should do (e.g., high vaccination rates) and the defects it should avoid (e.g., wrong-side surgery). In some cases, it can be measured objectively (e.g., percent of lab results reported to the patient within five business days), but subjective measures matter, too (e.g., level of patient satisfaction with the appointment-making process).

There is wide agreement that quality measurement and feedback serve essential roles in quality improvement. But quality measures should be considered carefully, since methodologic challenges can make them imprecise or misleading. For example, studies of health IT and physician adherence to treatment guidelines should, but do not always, measure how well users were trained and whether they actually used the system. Similarly, having a documented plan of care in the patient record does not mean the patient understood it, that it was fully appropriate for the patient, or that it was followed. Sometimes different subgroup outcomes mixed together in aggregate metrics are difficult to tease apart, such as breast cancer patients who are only somewhat responsive to a treatment such as Herceptin, because patients with different genetic subtypes were combined.[4] Measuring repeatedly over time is often the best way to capture meaningful changes in quality, such as when monitoring weight loss or blood sugar control.

Quality is becoming an even greater focus as medical practitioners and patients face more decisions that must be made in less time. Advances in personalized medicine will bring more molecular markers to test and consider during diagnosis, treatment, and care planning with each patient. Patients and providers may need extra time to make informed decisions and to keep up with new knowledge, fueling the demand for enhanced provider and patient decision support tools. Visit time may be further overloaded as practice demands increase, sicker patients are discharged from hospitals earlier, and more complex care is provided to outpatients. With accountable care contracts, providers are assuming greater responsibility and risk for the resource usage of a panel of patients, so quality, cost, and the patient experience will receive even more focus.

QUALITY AND PERSONAL HEALTH IT FROM THE PROVIDER PERSPECTIVE

What Quality Challenges Do Providers Face?

Most patients and many providers are unaware of all the quality gaps documented in healthcare. Health services research has shown that patients nationally receive only 54% of the preventive, acute, and chronic care they should receive, with levels of quality in specific medical conditions ranging from 79% for senile cataracts to 11% for alcohol treatment.[5]

The interactions among the patient, health care provider, and health care system depicted in Figure 5-1 are those that can have a negative effect on the patient's ability to follow a medication regime.

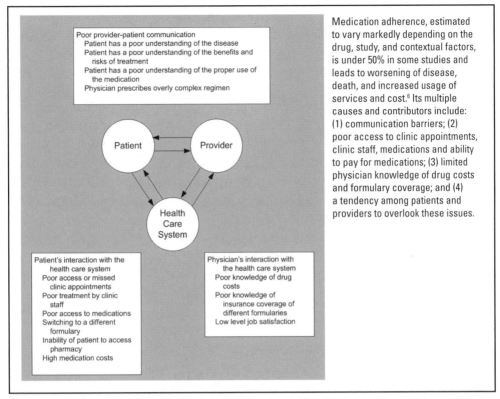

Poor provider-patient communication
Patient has a poor understanding of the disease
Patient has a poor understanding of the benefits and
risks of treatment
Patient has a poor understanding of the proper use of
the medication
Physician prescribes overly complex regimen

Patient

Provider

Health
Care
System

Patient's interaction with the
health care system
Poor access or missed
clinic appointments
Poor treatment by clinic
staff
Poor access to medications
Switching to a different
formulary
Inability of patient to access
pharmacy
High medication costs

Physician's interaction with
the health care system
Poor knowledge of drug
costs
Poor knowledge of
insurance coverage of
different formularies
Low level job satisfaction

Medication adherence, estimated to vary markedly depending on the drug, study, and contextual factors, is under 50% in some studies and leads to worsening of disease, death, and increased usage of services and cost.[6] Its multiple causes and contributors include: (1) communication barriers; (2) poor access to clinic appointments, clinic staff, medications and ability to pay for medications; (3) limited physician knowledge of drug costs and formulary coverage; and (4) a tendency among patients and providers to overlook these issues.

Figure 5-1. Barriers to medication adherence (Adapted from Osterberg L, Blaschke T. Adherence to medication. *N Engl J Med.* 2005; 353(5):487-497.)

Inadequate follow-up of lab results is another trouble spot that is recognized in health services research but not always apparent to providers or patients, and not easy to address. Physicians vary in their reporting of lab test results to patients. In one survey, up to one third of physicians reported they did not routinely notify patients of abnormal results.[7] Unless all results are reviewed and reported to the patient, it may be difficult to identify whether the patient had the test done in the first place, had a test that was mislabeled, or had test results that were not communicated back to the ordering physician. Sometimes administrative failures occur, such as being unable to identify a responsible physician for a test, test results that were misfiled or misplaced, a message left for a patient but never acknowledged, or some other communication breakdown with patients or others.

Among physicians surveyed in 2002, only 41% reported being at least somewhat satisfied with the way they managed test results, and 83% reported at least one delay in managing test results in the past two months.[8] Fewer than half of physicians reported having a mechanism to ensure that a patient with a marginally abnormal mammogram received a follow-up mammogram within six months. More than one third did not have someone else in their office whom they relied on to screen incoming results for abnormalities.

Problems with missed and delayed diagnoses are not uncommon. A study of 181 closed malpractice claims was conducted to understand root causes of missed and delayed diagnoses in ambulatory practice using data from four insurers.[9] The study found that harm to patients was serious in the majority of cases (59%), that cancer was the diagnosis missed most often (59%), and that system breakdowns most commonly involved failure to order an appropriate diagnostic test (55%), create an appropriate follow-up plan (45%), obtain an adequate history (42%), perform an adequate physical exam (42%), or correctly interpret a diagnostic test (37%). Diagnostic errors almost always resulted from multiple breakdowns rather than a single mistake—typical of complex system failures. Nearly half of all cases had an inadequate follow-up plan, including poor documentation, scheduling problems, and miscommunication among providers. This suggests that patients who ask for a follow-up plan and help to ensure its completion can make a difference.

Further analysis of a subset of 56 cases involving missed cancer diagnoses[10] identified possible prevention strategies. The error would have been prevented in 48% of cases given assistance from a non-clinical but meticulous patient advocate during the diagnostic process; in 66% of cases given adoption of diagnostic guidelines for breast and colorectal cancer; and in 35% of cases given improvements in communication among providers or between provider and patient. These data suggest that communicating with patients routinely would help reduce diagnostic errors.

Many more studies document quality gaps in different areas. EHR data is consistently found to be missing or error-prone. A 2004 study by Kaboli et al found almost 95% of medication lists had some inaccuracies.[11] Omissions (medicines taken by the patient but not listed in the EHR) were 25%; commissions (medicines not taken by the patient but listed in the EHR) were 12%. These results highlight the importance of medication reconciliation at each patient visit, as well as the limitations of clinical decision support in accurately processing drug-drug and other types of interactions that rely on accurate data. One third of patients had errors in their allergy and adverse drug reaction list—mostly omissions. Schnipper found gaps not only in medication information, but family history data as well.[12] Putting medication lists and medication allergies online can help here too. Some practices are engaging patients in reconciling their medications.[13] Patient-facing drug-drug interaction checkers can help, too.

Another study looked at medication changes upon hospital discharge compared to admission and found that 24% of medications that were stopped or redosed were suspected provider errors. They also found no patient understanding of 69% of redosed medications, 62% of new medications, and 82% of stopped medications.[14]

Frequently, limited care coordination and access to patient data lead to poor quality, as evidenced by tests that were unnecessarily repeated and uninformed decision making—often for patients seen by a variety of specialists without the benefit of a generalist to help connect the data and care activities.

Usability, "the effectiveness, efficiency and satisfaction with which specific users can achieve a specific set of tasks in a particular environment,"[15] is particularly challenging in EHR systems due to their complexity. EHRs support a large number of tasks, a large and diverse group of individuals, mobile physicians with high levels of distraction, and a variety of workflows for individuals as well as teams. Since poor usability can create cognitive strain among users and may contribute to patient risk, it is an extremely important area to address. The design of patient-facing systems also matters, as elegant design can change impenetrable data into understandable information.

Providers also face an explosion of knowledge, which presents a quality problem—how do they keep up? How do they set realistic expectations among their patients about what they know, and don't know? How should providers encourage patient learning amidst the volume of information and misinformation available on the Internet?

Since most of patients' care is provided in the community in which they live, a community-based health information exchange (HIE) or health record (e.g., health record bank), and a third-party personal health record (PHR) can also be used to bring externally-available data to both the provider's and the patient's attention. Likewise, specialists, hospitals, urgent care facilities, and others should feel responsible to "push" any relevant and important new information developed during the care they provided back to the primary care physician. The "Direct" secure messaging protocol can be used for this purpose and can even be used to simultaneously "cc" the patient's PHR.

How Does Patient Use of Personal Health IT Help to Address the Quality Challenges That a Provider Faces?

The quality problems just described are wide-ranging and significant. Although not simple to solve, there are many opportunities for the patient to help, especially through using personal health IT. Patients can play an important role in improving medication adherence, following up on lab test results, avoiding missed or delayed diagnoses, reducing EHR data errors, understanding discharge medications, and improving information sharing during care transitions and during care coordination in general.

Patient use of an advanced patient portal offered by his or her provider can address a number of quality issues that concern providers. By offering three main components—chart transparency, secure messaging, and in-context reference information, the patient portal brings patients into the process of information sharing, communication, and learning, as described in Chapter 4.

Many medication adherence challenges such as understanding medications, using them consistently, making adjustments when needed, and addressing barriers to use such as side effects, economic hardship, and literacy challenges can be addressed in part using personal health IT. For example, current medications visible in the patient portal can help remind the patient and caregivers of the intended regimen, or if there is a mismatch with what they expected to see, trigger a clarification call or message. Understanding about what to expect when taking a medication, when to contact a pharmacist or doctor about a side effect, and safety information such as drug interactions can be supported by educational handouts linked to the medication list or searchable in the portal or on the Internet.

Some advanced patient portals offer online tools to help patients reconcile the medications they are taking with what is listed, since it is common for medication regimens to change and for medication lists to be out of date.

The growing practice of sharing lab test results using a patient portal can close the loop on many failures in communication. Patients expecting a lab result to appear on their patient portal can alert someone if it does not. Patients can review their provider's interpretation of the lab results online and verify their understanding of what was discussed during a visit or phone call. Patients should be encouraged to review the details of lab and radiology reports themselves, since anecdotal reports suggest patient access to this information helps detect abnormalities that might have been missed by a busy physician. Patient review can also improve action-taking. In a study of patients with abnormal mammograms, patient understanding of the need for follow-up was independently associated with the delivery of appropriate care,[16] as further discussed in Chapter 4.

Patient access to laboratory results and reports has been used to streamline practice workload and workflow, replacing printed/mailed "Results Letters" with online copies that are available immediately. Besides reducing the risk of orphaned results that never find their way to the patient, this feature can help verify that a patient has read a Results Letter and its recommendations. By incorporating context-sensitive links for each lab test, such as those available from LabTestsOnline, patients not only see their result but can read background material about each test, the test sample, the reasons for the test, the result, and the implications of the result.

Many of the research findings concerning missed and delayed diagnoses identify communication errors and information sharing problems as root causes. Patients and providers who assume that health information will be made available and shared extensively, as a baseline expectation, reduce the likelihood of these kinds of problems.

Patient portals are critical technologies. Just as important are practice policies that ensure office staff will engage actively in information sharing and promoting these services to patients. The technology alone is not enough—people must develop the habit of using it. This is a challenge—and opportunity—for physician leadership.

Patient access to EHR chart information shared within the patient portal offers patients the opportunity to review past information when getting ready for a clinic visit or phone call, to check the accuracy of safety-related information such as the medications and allergies list, and if desirable, to copy chart information into a third-party system such as Microsoft Healthvault or another PHR system. Access to clinical data also enables patients to "know their numbers," gaining a deeper understanding of their health, health risks, or conditions, and thus able to participate more in partnership with their providers. Improving diagnostic accuracy and EHR data accuracy is possible by exposing data transparently with the patient, who is often able to identify errors or missing information.

Research with patients who use an eJournal as part of patient portal suggests a trend toward improved EHR data accuracy, a reduction in potential adverse drug events,[17] and adherence to prevention guidelines.[18] For clinical care purposes, having accurate data in the EHR matters most when the data are used for clinical decision making. Finding and reviewing information in the EHR record that is likely to be used in the future should be the main focus of the patient and the provider—such as medications, allergies, lab results, diagnostic reports, family history, health maintenance information (e.g., vaccinations), and past treatments that have worked or failed to work as expected.

Another way personal health technologies help is through communication tools within the patient portal that allow interactive or form-based requests from patients to the practice when scheduling or managing appointments, requesting prescription renewals, or requesting referrals. Providers benefit in several ways when patients use these kinds of tools. Each form-based request is a "perfect" request—it has all the information needed for processing. The vast majority of these requests replace a phone call with a quickly read, self-documenting message that can be recorded in the patient record. Patients also are often more thoughtful when they compose a message to their provider than when speaking on the telephone. This saves time and improves the provider's ability to close an issue soon after it was opened. When a message contains patient-supplied data such as a medication list critique or recent home blood pressure readings, it streamlines the conversation between the physician and patient while providing data that was not easily available prior to the use of this tool.

There is early evidence suggesting that patient use of portals can help providers meet quality measures. A 2010 Kaiser Permanente study showed that patients with chronic illnesses who used patient/physician email had higher HEDIS scores than a control group of similar patients. The study found that, "For patients with diabetes and hypertension, the use of secure patient-physician e-mail was associated with an increased likelihood that patients would meet each of nine HEDIS measures. In addition, when compared to matched controls, the use of e-mail was associated with a 2.0–6.5 percentage-point improvement in HEDIS performance.[19]

The Agency for Healthcare Research and Quality (AHRQ) published a literature survey in the summer of 2012 that showed promising results: "Substantial evidence exists confirming that health IT applications with patient-centered care-related components have a positive effect on healthcare outcomes."[20]

In a study published in 2012, funded by AHRQ, researchers found that patients who used interactive health records were almost twice as likely to become up-to-date on recommended preventive care. That included screening tests for breast, colon, and cervical cancers, and immunizations like the yearly flu shot. After 16 months, 25% of patients who used the online records were up-to-date on their preventive care—which was double the rate of non-users.[21]

Patients are increasingly using Internet resources to become experts in a health area of interest to themselves or someone close to them. By asking patients about what they find, their sources of information, and the tools they have found useful, providers can learn what is valuable and add important information the patient may not have considered, such as making sure the sponsorship of the website does not create a potential conflict of interest. One resource that compares patient data is the Internet site PatientsLikeMe, which offers patients tools for sharing symptoms, treatments, and experiences over time with other patients and families as they manage a condition. Chapter 9 covers in greater detail patient and provider use of online information resources, including communities and social media.

QUALITY AND PERSONAL HEALTH IT FROM THE PATIENT PERSPECTIVE

What Quality Challenges Do Patients Face?

Many of the quality challenges faced by patients overlap those faced by the provider, but with important differences. For typical patients, their lack of formal medical training, limited medical knowledge, lack of an insider network, and limited understanding of clinical care processes may leave them feeling vulnerable or largely unaware. They may also experience a deeper emotional connection to the health concern than the provider, especially if it is serious, frightening, or time-urgent.

Whereas a provider is usually skilled at integrating patient data, medical knowledge, process-of-care information, and patient preferences, patients do not usually have these skills developed. However, there are a number of things patients can do right away using personal health IT to impact quality. The Center for Advancing Health's Engagement Framework describes many actions patients can take[20] to participate knowledgeably and effectively in care. While the framework is not IT-specific, many of the actions it describes are supported by personal health technologies. For example, "bring a list of all current medications" is more easily accomplished by patients who use personal health IT with access to their medication list. Patients are also advised to prepare a list of question/issues in advance of a visit. An electronic patient portal allows the patient to share those questions in advance with their provider. This resource details 42 engagement behaviors and is well worth a close look by patients who want to take an active role in their care. It is also important for patients to consider that even a skilled provider may lack the time or the specific experience to help in all situations, so they should ask their providers what to do if that situation should arise.

Researching online medical knowledge to learn about the health concern with or without the assistance of a librarian or a health coach is an important way to begin learning. There are usually books, tapes, and increasingly, DVDs, websites, blog postings, and online communities and videos that can help patients learn at a comfortable pace and level (these tools are also discussed in Chapters 4 and 9). Patients will want to focus on learning about: (a) the diagnostic possibilities and how the diagnosis will be confirmed; (b) the treatment options, including risks and benefits for each even if an option is not offered by the provider offering consultation; and (c) prognosis, including implications for family members and social contacts.

Even if patients prefer that their clinical data and medical decisions remain largely the responsibility of the treating provider, it is important for them to establish access to as much data as possible for several reasons. First, without data transparency, incorrect or missing data will take longer to identify and could delay appropriate care or lead to undesirable effects. For example, incomplete drug allergy information might mean a contraindicated drug is prescribed. Or missing lab tests could slow the adjustment of medication that is being titrated to a target blood level. Patients can also look for resources to help them learn how to recognize quality in the course of care, the defects that are most common for patients in a similar situation, and the actions they can take to manage those situations.

While some studies have shown a correlation between various kinds of personal health IT and higher levels of patient engagement, it is unclear whether the use of personal health IT caused the engagement or resulted from it. Personal health IT tools cannot force patient engagement, nor can they build trust by themselves; they can, however, present an opportunity for patients to engage by helping them "know their numbers" and become more knowledgeable about their health or their conditions. They can support patients in taking the first steps in engagement or move deeply into personal empowerment and self-actualization. Gamification, which will be discussed in both Chapters 9 and 12, can be used to make the use of personal health IT tools more interesting and even fun for patients.

A strong reason for patient vigilance—looking at clinical test results, medication list, allergy list, and problem list, for example—is that no single provider can attend to all the information and needs of each of his or her patients. Chances are high that some errors will be identified and corrections requested, although typically only a few of those errors would significantly improve the care received by the patient.

Patients with language, visual, auditory, health literacy, or other kinds of limitations should be offered assistance using delegates or alternative ways of communicating with the care team. And beyond the challenges outlined so far, patients have to remain flexible in their decision making—knowing when to accept a treatment or diagnostic plan that represents their wishes, and when to challenge the wisdom or recommendation of a provider or someone with expertise. Having a good care experience despite fragmented information, communication, and incentives, requires patients to be constantly learning—what works, what does not, what process step comes next, and what other resources are available. Learning to understand risk statements, or having "statistical literacy" is an increasingly important skill for understanding health concepts.[21] Perhaps social media and online information tools can help patients to conduct healthcare research such as new gene testing for a condition that runs in their families, or to find people with similar clinical data and concerns who are open to discussions.

How Does Patient Use of Personal Health IT Help to Address the Quality Challenges That a Patient Faces?

Patients can use personal health IT such as a basic patient portal to stay organized, access information, communicate effectively and efficiently, remember things that might otherwise be forgotten, collect data, notice when providers have missing or wrong information, learn how to navigate the health system, and be on an equal footing with the care team.

What Should a Provider Do to Encourage Patient Use of Personal Health IT for Quality?

Encouraging patient engagement and patient use of personal health IT requires a multi-level approach. Setting overall expectations with practice providers and staff to adopt a set of core principles will help establish a consistent, practice-wide approach of encouraging patients to participate in the use of technology for care and a culture of inclusion. The first principle is to adopt a service approach and mindset—what will delight the patient?—and practice this with every patient contact.

The second principle is offering transparency and access to clinical information as an essential component of care delivery. Offering patients equal access to information, even if they choose not to use it, sets an expectation of participation. Access should include not just the data but the tools to manipulate the data, such as graphing lab results or selecting information extracts for Blue Button download (see Chapter 12 for more information on Blue Button). Access also means it is easy for patients to fully authenticate to the portal without delay, even if recovering a lost username/password. Access also means introducing the shortest embargo periods between lab results being available to providers and availability to patients (these are ideally negotiated between patient and provider if immediate access is not granted), and the most expansive information access possible including the result, any/all data that would provide context for understanding the result, any quality information to help substantiate the results accuracy or understand uncertainty associated with it, and contact information for access-related help or feedback on results themselves. Every staff member from physicians to front desk to phone contacts should be equipped to help patients obtain access to their data and related information, or to direct them to someone who can help immediately.

The third principle is approaching each patient as a care partner. This means anticipating that patients will play an active role in their care and offering guidance/coaching to encourage that. Each patient is signed up for patient portal, uses it as part of the visit routine, receives lab results and communications via the portal ("that's how we provide care, here…"), and is asked about his/her preferences for communication, information sharing, decision making, and coordination, such as who should be involved (if anyone) besides themselves, and adding these data to the profile information for the patient—as part of an electronic system if possible.

The fourth principle is to focus on patients' user experience, gathering their viewpoint informally and formally (through surveys), envisioning how they will navigate the care process—including technology, their level of understanding and engagement, and taking a broad (beyond the practice) perspective.

At a tactical level, there are several mechanisms that will help to accelerate practice and patient engagement with patient portals:

- First, as soon as a go-live date is set, begin collecting email addresses from patients who visit or have telephone contact with the office. When the go-live date arrives, email everyone with the link to the enrollment site.

- Second, offer instant authentication to avoid any delay in processing enrollment requests or distributing passwords. Offer this feature during password recovery as well—it drives up portal use significantly.

- Third, ensure that patients who send a message to the practice receive an immediate reply that makes clear when their request will be completed.

- Fourth, engage providers and practice staff (everyone) in communicating about the service to all patients.

CONCLUSION

Why is patient engagement using personal health IT tools good for quality, for patients?
- Enables patients to better understand their own health, facilitating informed decisions.
- Provides patients with convenient ways of accomplishing their healthcare tasks.
- Provides patients with effective ways to interact with their providers.
- Provides patients with ways to track and understand their health and health conditions.

Why is patient engagement using personal health IT tools good for quality, for providers?
- Helps providers to keep their patients healthy by educating and informing them.
- Helps providers treat patients effectively when they are sick, using the right level of care.
- Helps providers by enlisting the patient in spotting potential mistakes, avoiding disruptions in care, and ensuring that information is shared during care transitions.
- Delights patients, increasing loyalty.
- May reduce administrative costs if personal health IT supports workflow well and the volume of use is sufficient.
- May increase health status and HEDIS scores for chronically ill patients.

Why is patient engagement using personal health IT tools good for quality, for the system overall?
- Patients are an undertapped resource; many will be vigilant in finding opportunities for quality improvements.
- May reduce cost or limit rising costs in the system.
- May reduce waste and redundant activities.
- Many opportunities for feedback and system learning at an individual or provider practice level can have an impact on the system as a whole.

For more information, read the following case studies at the end of this book:

Case Study 4: From the Heart: The Intersection Between Personal Health and Technology

Case Study 7: Everyone Performs Better When They Are Informed Better

Case Study 9: Restoring Trust with Information

Case Study 11: The Power of Information and the Partnership of Trust

Case Study 13: Meaningful Use Implementation and Patient Activation

Case Study 14: Moving the Point of Care into Patients' Daily Lives via Text Reminders

REFERENCES

1. Kohn LT, Corrigan J, Donaldson MS. *To Err is Human: Building a Safer Health System.* Vol 6: Joseph Henry Press; 2000.

2. Institute of Medicine CoQoHCiA. *Crossing the Quality Chasm: A New Health System for the 21st Century.* Washington, DC: National Academies Press; 2001.

3. Caper P. Defining quality in medical care. *Health Aff (Millwood).* 1988; 7(1):49-61.

4. Sweet KM, Michaelis RC. T*he Busy Physicians Guide to Genetics, Genomics and Personalized Medicine, First Edition.* Springer; 2011.

5. McGlynn EA, Asch SM, Adams J, et al. The quality of health care delivered to adults in the United States. *N Engl J Med.* 2003; 348(26):2635-2645.

6. Osterberg L, Blaschke T. Adherence to medication. *N Engl J Med.* 2005; 353(5):487-497.

7. Boohaker EA, Ward RE, Uman JE, McCarthy BD. Patient notification and follow-up of abnormal test results: a physician survey. *Arch Intern Med.* 1996; 156(3):327.

8. Poon EG, Gandhi TK, Sequist TD, et al. I wish I had seen this test result earlier!: Dissatisfaction with test result management systems in primary care. *Arch Intern Med.* 2004; 164(20):2223.

9. Gandhi TK, Kachalia A, Thomas EJ, et al. Missed and delayed diagnoses in the ambulatory setting: A study of closed malpractice claims. *Ann Intern Med.* 2006; 145(7):488.

10. Poon EG, Kachalia A, Puopolo AL, et al. Cognitive errors and logistical breakdowns contributing to missed and delayed diagnoses of breast and colorectal cancers: A process analysis of closed malpractice claims. *J Gen Intern Med.* 2012:1-8.

11. Kaboli PJ, McClimon BJ, Hoth AB, et al. Assessing the accuracy of computerized medication histories. *Am J Manag Care.* 2004; 10(11 Pt 2):872.

12. Schnipper J. Where errors occur: inadequate family and medication histories. *Forum.* 2007; 25:10-12. http://www.rmf.harvard.edu/~/media/Files/_Global/KC/PDFs/v25n2whereerrorsoccur.pdf. Accessed December 20, 2012.

13. Dullabh P. Policy Issues Related to Patient Reported Data. In: Health and Human Services, Office of the National Coordinator for Health Information Technology, ed. Washington, DC: Health and Human Services, Office of the National Coordinator for Health Information Technology; 2012.

14. Ziaeian B, Araujo KLB, Van Ness PH, et al. Medication reconciliation accuracy and patient understanding of intended medication changes on hospital discharge. *J Gen Intern Med.* 2012:1-8.

15. HIMSS EHR Usability Task Force. Defining and testing EMR usability: Principles and proposed methods of EMR usability evaluation and rating. 2009. http://www.himss.org/content/files/himss_definingandtestingemrusability.pdf. Accessed December 20, 2012.

16. Poon EG, Haas JS, Louise Puopolo A, et al. Communication factors in the follow-up of abnormal mammograms. *J Gen Intern Med.* 2004; 19(4):316-323.

17. Schnipper JL, Gandhi TK, Wald JS, et al. Effects of an online personal health record on medication accuracy and safety: a cluster-randomized trial. J Am Med Inform Assoc. 2012; 19:728-734.

18. Wright A, Poon EG, Wald J, et al. Randomized controlled trial of health maintenance reminders provided directly to patients through an electronic PHR. *J Gen Intern Med.* 2012; 27(1):85-92.

19. Zhou YY, Kanter MH, Wang JJ, et al. Improved quality at Kaiser Permanente through e-mail between physicians and patients. *Health Aff (Millwood).* 2010; 29(7):1370-1375.

20. Finkelstein J, Knight A, Marinopoulos S, et al. Enabling Patient-Centered Care Through Health Information Technology. Agency for Healthcare Research and Quality; 2012. http://www.ahrq.gov/clinic/tp/pcchittp.htm. Accessed December, 2012.

21. Krist AH, Woolf SH, Rothemich SF, et al. Interactive preventive health record to enhance delivery of recommended care: A randomized trial. *Ann Fam Med.* 2012; 10(4):312-319.

Chapter 6

Patient-Provider Communications: Communication Is the Most Important Medical Instrument

By Dave Chase

"A good scalpel makes a better surgeon.
Good communication makes a better doctor."

−Josh Umbehr, MD

INTRODUCTION

The quantum improvement in the depth and breadth of communication seen in the consumer Internet is just now beginning to have a strong impact on healthcare. Tools that are commonplace in a consumer context such as smartphones and text messaging are now becoming accepted into the healthcare enterprise. Just as we have seen an array of new communication applications on our smartphones, so too will new healthcare-specific communication tools emerge to respond to the changing expectations of consumers. A wide array of communication tools are available ranging from simple secure messaging to text messages, video (live and asynchronous), and much more. In an email exchange, Wendy Sue Swanson, MD, MBE, FAAP, stated the following: "I don't think you can overstate the importance of communication in clinical care. Even with devices, robotics, genomics and personalized care−it all rests, and depends on, clear communication."

PATIENTS ARE THE MOST IMPORTANT MEMBER OF THE CARE TEAM

It has long been said that the most important member of the care team is the patient (or the patient's family members). Quite simply, in a world where providers are compensated on quality and outcomes, it is nearly impossible to run a successful practice without building it around a patient-centered approach. Clinicians can provide guidance and coaching, but ultimately it is the patient who navigates the way back to full health. Unfortunately, too often, patients get "lost" on that journey back to health.

Consider Figure 6-1. Healthcare providers control most decisions that drive outcomes in emergency or high-acuity cases, the most obvious example being the care of an unconscious patient in the hospital. In contrast, in low-acuity situations such as managing a chronic condition, patients and/or their families are clearly in control of the actions that will drive outcomes. Whether adhering to exercise, diet, or prescription plans, or recognizing and acting promptly on symptoms of relapse or adverse drug effects, patients and families play the central role in determining the results. (Note: While it is true that in acute or emergency situations there often is not sufficient time to take the family into account, this is not intended to suggest that there is no role for patient and family decision making in high-acuity cases.)

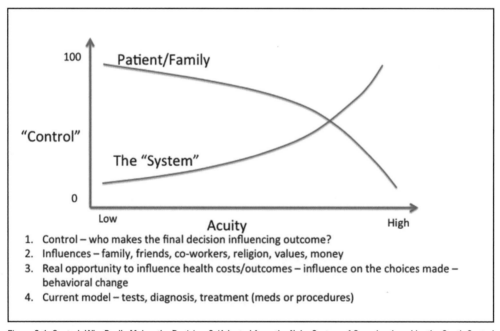

1. Control – who makes the final decision influencing outcome?
2. Influences – family, friends, co-workers, religion, values, money
3. Real opportunity to influence health costs/outcomes – influence on the choices made – behavioral change
4. Current model – tests, diagnosis, treatment (meds or procedures)

Figure 6-1. Control: Who Really Makes the Decisions? (Adapted from the Nuka System of Care developed by the South Central Foundation of Alaska.[1])

The importance of patients' role is clear, given that 75% of healthcare spending results from chronic conditions. Poor decisions made while a condition is in a low-acuity state can rapidly lead to high-acuity flare-ups that drive large medical bills. As Dr. Wendy Sue Swanson stated, "The steering wheel should be attended by the patient." After all, 99+% of patients' lives are spent away from healthcare providers, and it is patients who are in the driver's seat.

The Need for a "Healthcare GPS"

Today, for many patients, their experience in the healthcare system resembles sending them to a foreign land, giving them directions scrawled in a second language, and pushing them out the door to find a destination called "health." We tell patients that if they get lost, they should come back and we will give them directions again for the price and time of another appointment. Most patients just muddle through on their own, not wanting to appear ignorant, not wanting to bother their providers, or finding communication with their providers a troublesome task.

It is well understood that patients forget or misunderstand more than 80% of the information their doctors tell them during a visit. Using their own employees, IBM studied patient retention of instructions given by their physicians. The employees forgot more than 60% of what they were told within one day despite being healthier and smarter than the average citizen. What is needed instead is the equivalent of a global positioning system (GPS) for healthcare. Like a GPS, it would know where a patient is on the journey back to full health. The "GPS" in this analogy represents a variety of personal health communication and tracking tools. For example, a patient with congestive heart failure (CHR) might step on a scale daily and send that information wirelessly back to his or her provider. If the recorded weight indicates that the patient is off-track and a problem is developing, that patient might receive suggestions via email or text, or a phone call from a care manager. Effective, timely communication can act like a GPS device to get the patient back on course.

In the current system, many factors conspire to create the reality of patients who feel disconnected from their doctors, unsure what actions they should take for their own health, seeing individual providers who do not talk to one another or act in a coherent, coordinated manner. Historical reimbursement models have contributed to this dynamic, both by rewarding piece-work and because there are no incentives for coordinating care. Patients have not had clear ways to determine which providers will do a better job of communicating with them or building relationships, and have few ways to ask questions or build relationship between office visits. Considering that people retain less than 20% of what a doctor tells them, this lack of communication and patient retention is a brutal combination driving sub-optimal outcomes and "lost" patients.

Email Is Where It Begins

Secure email remains the "killer app" that is the starting point for most patient portals. Though it is an old technology, it can make a significant difference in building trusted relationships between providers and patients in any setting—and may be even more important for establishing a sense of direct connection in larger settings, where more layers of process inevitably make it harder to feel a direct connection with your doctor.

In a study[2] of 35,423 people with diabetes, hypertension, or both, the use of secure patient-physician e-mail within a two-month period was associated with a statistically significant improvement in effectiveness of care, as measured by the Healthcare Effectiveness Data and Information Set (HEDIS). In addition, the use of e-mail was associated with an improvement of 2.0–6.5 percentage points in performance on other HEDIS measures such as glycemic (HbA1c), cholesterol, and blood pressure screening and control.

Recently published data indicate that Kaiser patients enrolled in their patient portal, which includes secure messaging with doctors, access to clinical data, and self-service transactions, are 2.6 times more likely to stay with the organization than those who do not participate online (see more on Avoiding System Leakage below).[3] Countries such as Denmark provide incentives for doctors to communicate electronically. The result: 80% of physician/patient communication in Denmark is asynchronous[4] (i.e., people talking to each other serially rather than simultaneously). At first, that can sound high until we think about the rest of our lives, whether it is conducting business or communicating with friends, where asynchronous communications (e.g., such as email, voicemail, or texting) are the norm.

Email can be one way to address the problem that patients remember so little of what they are told in the provider's office. Other options include providing a clinical summary to patients after each visit (as recommended by Meaningful Use measures) and providing documentation of a care plan online for patients to refer to later.

INNOVATIVE CARE MODELS AND PHYSICIANS POINT THE WAY TO THE FUTURE

Organizations such as Kaiser Permanente and the Cleveland Clinic rightfully are held up as leaders in physician-patient communication because of the way they have incorporated secure messaging, access to clinical information, educational resources and other online tools into their practice environments. However, not every organization has the resources of a Kaiser Permanente or Cleveland Clinic. This section focuses on low cost ways to enhance communication between providers and patients, and highlights two types of examples:

- Individual physicians who have used low cost and free tools to improve communications.

- Low-price primary care practices built on strong communications that have been successful in achieving the Triple Aim: improving the individual experience of care; improving the health of populations; and reducing the per capita costs of care for populations.

Individual Doctors Use Low-Cost, High-Impact Communications

Even before new financial models fully take hold, many passionate doctors want the best possible outcomes for their patients. Some have created their own content—frequently using YouTube—in order to provide education that really fits their needs. What may seem like a one-way communication tool becomes two-way if covering the basics in the video enables the patient and provider to have a deeper conversation in the office or via email. Some patients even ask general questions in the comments on the video. In the process, the doctors who produce the videos also get the residual benefit of marketing since some of these doctors get 20% of their new patients from social media and videos. Each of the examples below was created using a free tool without any special technology.

Perhaps because Dr. Wendy Sue Swanson was a teacher before she was a pediatrician, she was naturally drawn to using videos and blogs to educate her patients and their families.[5] Topics range from "Understanding Growth Charts" to "Vaccines," as well as other topics that she is frequently asked about. Dr. Swanson shared the reaction she gets from patients: "I'll launch into something in clinic and a family will say, 'Dr. Swanson, you don't have to explain that, I read your blog post/saw your video, etc!' As a result, we start at a different place. A place that feels easier to connect, more informed, and one with more respect for our mutual vantage points."

Ryan Neuhofel, MD, ("Dr. Neu") uses video to explain the Hemoglobin A1c lab test using an M&M's metaphor, highlighting why the test is so important when monitoring diabetes.[6] Many physician-created videos are simply talking heads replicating what the doctor would otherwise say in a face-to-face encounter. In this example, Dr. Neu has simply recorded what he might otherwise sketch on a piece of paper during an office visit. Despite the fact that the video does not use special effects, the patient has the sense of a personal encounter with the doctor. The benefit for Dr. Neu is that instead of taking valuable time in each office visit to explain this test, he can record it once and direct patients to it. Patients are able to consume the video on their terms, even if it means playing it a few times to fully understand it without feeling uncomfortable about making the doctor repeat his explanation.

Natasha Burgert, MD, described the benefit to her practice and patients of using video as follows:

"Investing time in relevant and complete posts actually saves me time in the long run. Questions I am repeatedly asked, like 'How do I start solid foods?' can be answered quickly and completely by directing them to my site. This saves face-to-face clinic time for more specific concerns for their child. I can actively communicate, acknowledge, and positively influence the choices that my families make for their children between checkups. My anticipatory guidance can be repeated, reinforced, and repeated again."[7]

The benefits of doctor-created videos are not limited to primary care. Doctors such as orthopedic surgeon Howard Luks also have been realizing the benefit of videos. As Dr. Luks stated, "Every patient who sees my videos prior to their visit says, 'OMG it's you, you're just like you are in the videos.' Very powerful. It humanizes your practice."[8] One example of the type of topic that might otherwise consume valuable time during an appointment that can be addressed via video is the question of why patients need to undress for part of the exam he conducts. He created a video that explains the medical need and puts his patients at ease for what might otherwise be an uncomfortable conversation.

PATIENT-PROVIDER COMMUNICATION NOT LIMITED TO PHYSICIANS

One benefit of technology is that it can extend the reach of finite and expert resources. For example, behavioral health professionals are not readily accessible in many locales. Automated tools, as well as asynchronous communications, can remove the dependency on face-to-face encounters. While face-to-face interaction is always desirable, it is not always feasible. The Case Studies in this book by DeeAnna Nagel and Eve Phillips expand on the opportunities for patient engagement using tools such as online training to teach cognitive behavioral therapy (CBT) and online therapy using tools such as email, chat, and video interactions.

LOW COST PRIMARY CARE

Beyond individual physicians, new practice models have emerged that have patient communication as a centerpiece. One high-growth practice model included in the ACA[9] is called Direct Primary Care Medical Homes (or DPC for short). DPC practices are very similar to the Patient-Centered Medical Home (PCMH) model, but they use a straight flat fee per month that is paid directly by the patient and avoids insurance bureaucracy for day-to-day healthcare. DPC practices use the "GPS for healthcare" approach (e.g., regular asynchronous communication via email, remote monitoring) and have achieved the highest patient satisfaction scores (e.g., scores higher than Apple or Google[10]) in very low-overhead practices. DPC practices that have been around long enough to publish outcomes have patient populations with greater than average percentage of people with chronic conditions. Like a micro accountable care organization, DPC practices get a fixed amount to keep their patients as healthy as possible. A few examples of documented results[11] include:

- 20.3% improvement of hypertensive patients with blood pressure under control;

- Average drop of 42 points in systolic blood pressure (SBP) for patients who enter with systolic blood pressure greater than 160;

- 47-63% smoking cessation rates for patients with diabetes, COPD (chronic obstructive pulmonary disease) and coronary disease;

- 19% reduction in sick days and 50% reduction in days not productive at work (aka presenteeism); and

- 40-80% reduction in hospitalizations, specialist referrals, and emergency department visits.

Enabling Technology

DPC practices are just one example of practices that put a premium on communication even if that simply means direct access on the phone. One advantage newer practices have is a clean-slate approach to technology. In contrast to large providers spending millions of dollars deploying patient portals, many newer practices use flexible and low-cost cloud-based software. Some have deployed their systems in just a day but then evolve and expand the use of their systems. These start small with the basics of a patient portal and progressively use more as they advance their patient communications practices using an array of tools including the following:

- Secure messaging between patient and providers;
- Sending patient labs results;
- Allowing for medication refill requests;
- Appointment requests;
- Two-way patient-provider mobile health tracking;
- Multi-provider patient portal that allows patients to keep all of their health information in one place rather than spread across multiple silos;
- Biometric device connection (e.g., wireless scales, activity monitors, blood pressure cuffs);

- Family/proxy capabilities so pre-approved family members can gain access;

- Patient-enabled scheduling—more than just a request, they can reserve an appointment—just like one can reserve a table at a restaurant;

- Practice/doctor website that is not an obvious templated site, but that provides valuable text and video content;

- One-too-many publishing of patient education information to cohorts of patients (e.g., COPD patients receive a stream of content related to treatment options and self-management tips); and

- Intake forms and condition diaries no longer have to be repeatedly filled out manually and then keyed in by a provider's staff. Patients can electronically fill out forms on their own schedule, eliminating a redundant task for staff.

Future EHR Requirements

In an article entitled "9 ways future EHRs need to support ACOs" in *Healthcare IT News*,[12] one of the leading thinkers on the future of health IT, Shahid Shah, was interviewed about future health IT needs that match up with the requirements outlined in the previous bulleted list. "The EHR systems and IT required for Meaningful Use (MU) is quite different from what will be required for ACOs," Shah said. "It will be nowhere as easy for existing legacy EHRs to simply retool their current platforms, like they did for MU [Meaningful Use]."

Shah went on to outline nine ways future EHRs need to support value/outcome-based models beginning with a focus on patient-provider communications which he listed as the #1 priority: "Sophisticated patient relationship management (PRM)." According to Shah, today's EHRs are more document management systems, rather than sophisticated, customer/patient relationship management systems. "For them to be really useful in ACO environments, they will need to support outreach, communication, patient engagement, and similar features we're more accustomed to seeing, from marketing automation systems than transactional systems."

ADVANTAGES OF IMPROVED COMMUNICATION

Increased Adherence to Care Plans

As stated earlier, patients (and their families) are central to determining the outcomes for people with chronic conditions. Tools such as the videos mentioned earlier help to ensure comprehension of one's diagnosis or care plan. This can be followed up by email and automated communication to track patients and remind them of things they should do. Ted Epperly, MD, FAAFP, explains the value of communication in adherence to care plans[13]: "A patient does not care how much his or her physician knows until they know how much the physician cares. Only after a trusted relationship is established and the patient knows that the physician is there for them in a caring, non-judgmental manner will they believe them and follow their advice, guidance, and recommendations. It is in this relationship that shared plans and mutual responsibilities occur, leading to a patient's compliance with his or her healthcare plan."

Improve Provider Efficiency and Coordination

If you were to observe virtually any doctor for a month, you could find that he or she has their own FAQ for various conditions, diseases, prescriptions, etc. They are essentially hitting the replay button hundreds of times a month during office visits. Smart doctors recognize that there is a better way. Patients and family members benefit greatly when doctors have a mini package of curated content (video, articles, etc.) that is developed for groups of patients (diabetics, parents of infants, etc.). This is predominantly a manual process today (e.g., writing down web addresses in an appointment or emailing them afterwards). Modern patient relationship management systems automate this process and allow patients to digest the content on their terms. For example, many patients are embarrassed to ask the doctor to repeat something they did not understand, so they walk away confused. This has been a boon to sites such as WebMD—patients fill gaps of information by going to "Dr. Google." Most clinicians realize that communication is the most important "medical instrument," yet time pressures do not allow them to spend a great deal of time with patients. Thus, they must come up with other ways to enable effective communications.

Some doctors identify content for their patients on their own but there are low- and no-cost services that take less than two hours of the physician's time to record a series of one- to three-minute video vignettes for the most commonly discussed items as noted earlier. While initially daunted by this, doctors quickly recognize that all they are doing is simply recording what they are already saying every day. Larger systems handle this with embedded connections to health encyclopedias that provide consistent and reliable content, as well as personalized content.

Virtually Extend Physician Time

Even though it is virtual, when doctors record short video vignettes, their patients feel like they have interacted with them. Further, they can digest the information on their terms. It essentially extends the appointment without costing the doctor more of his or her time. This new way for doctors to connect with patients enables them to change the nature of the face-to-face encounter, focusing more on individual care than repeating rote information hundreds of times.

Email systems are also ways for doctors to virtually extend their time. Lower-level staff may be able to triage emails and handle them without the physician's intervention. Doctors also may be able to move some activities from in-person visits to email—enabling them to focus on hands-on care for more acutely ill or difficult patients.

Avoid the Cost of Face-to-Face Appointments

Employers and patients are beginning to appreciate the high cost of face-to-face visits that are often unnecessary. In the traditional fee-for-service reimbursement model, the only way providers are paid is if they have face-to-face appointments. Doctors consistently point out that two thirds of their patient interactions do not require a face-to-face interaction; however, the reimbursement model demands office appointments in order to be paid. Don Berwick, MD, outlined the burden of face-to-face appointments in a visionary book The Commonwealth Fund published entitled *Escape Fire: Lessons for the Future of Health Care*[14]: "I believe that this new framework will gradually reveal that half or more of our encounters—maybe as many as 80 percent of them—are neither wanted by patients nor deeply believed in by professionals. The healthcare encounter as a face-to-face visit is a dinosaur. More exactly, it is a form of relationship of immense and irreplaceable value to a few of the people we seek to help, and these few have their access severely curtailed by the use of visits to meet the needs of many, whose needs could be better met through other kinds of encounters."

Let us do a simple calculation of the cost to the U.S. workforce to illustrate the point:

- In a *Health Affairs* study, 86.3 minutes of the 102.7 minutes involved in having a doctor's appointment is all about getting to and from a clinic and the accompanying waiting and hassles.

- Just over 150 million people are in the workforce, and the average wage equates to $22/hour.

- An average of 956 million medical appointments occur each year.[15] We will assume half of those are for people in the workforce, which would equate to 478 million medical appointments. Even if we assume only half of those face-to-face appointments could be replaced, that is still 239 million appointments.

- The calculation would be 86.3/60 x 239 million appointments x $22/hour = $7.6 billion of avoidable lost productivity per year if people do not waste time driving to and waiting for appointments.

- There are additional costs such as gas, parking, and tolls as well that are not included in that calculation.

Tools like email and online education that is provided in the doctor's voice or in a context vetted by the doctor clearly help avoid at least some visits for less acute conditions. A Kaiser Permanente study shows that phone and email were able to reduce visits by 25%.[16] When incentives are aligned, significant additional capacity exists in the healthcare system.

AVOIDING SYSTEM LEAKAGE

In an ACO, as well as in any capitated model of care that does not force people to receive care from one system, a top concern is leakage—when a patient within an ACO receives care outside of the system running the ACO.

Those in the field of consumer marketing know that it is critical that their brand is top-of-mind when a consumer is in the market to buy goods or services they can supply. In healthcare, the moment of truth is when a health event occurs. If the ACO provider has not established a trusted personal relationship with patients and made it relatively simple to receive care, it is highly unlikely they will be top-of-mind when a patient needs help. If some other provider is top-of-mind, patients are significantly more likely to go outside the ACO system for care. To inoculate against this kind of leakage, a consumer marketing mindset is critical but remains a foreign concept to most healthcare providers. Key aspects of a consumer marketing mindset applied to healthcare include:

- Building a trusted personal relationship;

- Making it easy for patients to communicate with physicians;

- Reaching out to patients when they are not sick with touchpoints or health reminders;

- Building convenient self-service capabilities into the physician portal; and

- Making it easier for the patient to get an appointment with a provider inside the ACO than outside of it.

STAFF EFFICIENCY

The traditional approach to healthcare largely ignores the central role patients (or families) play in driving outcomes and if continued, could prove to be a fatal flaw in the new reimbursement models. Throwing resources such as care coordinators at the problem can help, but it would be better to combine the best of human- and technology-driven communication methods to involve the patient in the process.

The following statement[17] is from a "Pioneer ACO" who described the importance of weaving patients into the process: "With our finite resources, we must figure out ways to offload what we have thought of as tasks that needed to be done by our staff. In most cases, it's the patient who can do it more effectively. In the process, the patient is more engaged and it's more efficient for everyone."

Myth and Reality of Physicians Overwhelmed by Emails

Physicians are understandably concerned about being overwhelmed by emails if they provide an option for secure messaging. As mentioned earlier, financial incentives have a big effect on physicians' willingness to take on what many perceive to be "more unpaid work." Interestingly, the physicians who have given out their phone number or enabled secure email (without remuneration) have not found they are overwhelmed by any means. In fact, they have experienced a number of benefits.

Ted Epperly, MD, who has been a family doctor for decades, described his experience as follows: "I give them [patients] both my phone number and a way to contact me via email. In 32 years of being a family physician I have had this privilege abused less than 5 times. On the flip side, it has led to many occasions where I have been able to expedite care and save countless number of office visits, ER visits, and hospitalizations. That is patient-centered care, and I personally feel better for it."

Howard Luks, MD, is an orthopedic surgeon who also has experienced similar benefits. "Physicians underestimate the fact that opening up a digital channel to facilitate post-visit, post-surgery, etc. comments and questions can and does provide a very real ROI [return on investment] if you dive into the typical workflow pattern that evolves when a patient calls with questions. If my assistant or nurse is tracking me down after fielding a phone call, they are not available to perform work that will lead to income. If I can answer a question with a brief email, it saves everyone time and enables him or her to remain active in meaningful tasks. So... there are tangible reasons why the use of digital communications in this day and age are worthwhile, but many are not savvy enough to realize the upsides and fear that they will be inundated with an enormous number of useless emails. I can tell you that it never happens, and in fact, patients start most every email with "Sorry, but I..." They are very respectful of the opportunity to engage in this format, and they are very cognizant of the fact that it does take away from my other clinical-related activities.

It is clear that providers can impact how their patients use secure messaging. Those providers who suggest that their patients follow up digitally, "After you've taken these new medicines for a couple of weeks, please send me a secure message and tell me how you are doing," and who advertise their willingness and ability to engage with patients via secure messaging will have more digital encounters than their counterparts who mention it rarely or not at all.

As providers do more of their visits via secure messaging, however, systems will need to think about new models for compensating providers that acknowledge that writing a thoughtful message to a patient does take time and needs to be balanced with other work. Some organizations expect over a quarter of their doctors' time will be spent responding to email.

IMPORTANCE OF ALIGNED FINANCIAL INCENTIVES

Importantly, the results highlighted in the DPC model discussed earlier and the case studies of Kaiser Permanente and Denmark use an economic model that aligns financial incentives with providing effective and efficient care, regardless of how that care is delivered. As a Congressional Budget Office analysis noted,[18] "How well health IT lives up to its potential depends in part on how effectively financial incentives can be realigned to encourage the optimal use of the technology's capabilities." Instead, the U.S. has a system that doesn't align financial incentives with health outcomes, and therefore, the results are sub-optimal.

Dr. Ted Epperly[19] stated: "It's like having the five best basketball players on the planet. According to our system of payment, we give each of them a basketball and tell them to dribble around and shoot at will. We tell them that every time they shoot the ball they will be paid. They don't even have to make the basket to get paid; they just need to shoot the ball. And so we have these five superstar basketball players dribbling and shooting constantly in a fragmented, non-integrated, non-coordinated fashion. And to make matters worse, they don't communicate with each other in any effective way to modify their behavior for each other's aid, let alone (extending the analogy to healthcare) their patients'. We then play a team from Spain, England, Canada, France, or Germany, and we get beat. Why? Not because we don't have the best basketball players on this planet; it is because we do not pass the ball and play as a team. When you step back and look at how our nation performs healthcare, we stack up poorly against the other industrialized nations in terms of outcomes. Not because we don't have talented people, but because we care for our patients using a fragmented, non-integrated, unaccountable, non-functioning system of record-keeping and communication. It is not connected well with information sharing."

Communication is the driver of satisfaction in any relationship, whether it is personal or professional. It becomes all the more important as Medicare is including patient satisfaction as a factor in reimbursement, as highlighted in a recent *New York Times* article[20]: "The ratings are based on Medicare-approved surveys, which hospitals hire companies to give to a random selection of patients after they are discharged. Some surveys are given by phone, others by mail. All ask the same questions: Did the doctors and nurses communicate well? Was pain well-controlled? Was the room clean and the hospital quiet at night? The surveys go to younger patients, as well as Medicare beneficiaries."

The first two questions on Patient Experience address communication:

• How well nurses communicated with patients; and

• How well doctors communicated with patients.

Whether out of desire or necessity, consumers are ready for improved communication so they can save on their healthcare costs. It is expected that roughly one third of the workforce will be permanent freelancers, contractors, consultants, etc., with little expectation of employer-provided insurance.[21] Even for those with employer-provided insurance, employees today are responsible for an ever-growing percentage of the premium and cost of care. The current average is that employees pay 30% of the insurance premium costs (up from 10% in the recent past) and may also be responsible for significant deductibles. Some speculate that 50% of the workforce will directly purchase their own healthcare via the health insurance exchanges (to be implemented in 2014 as a result of the federal health reform).

This increased onus on individuals coincides with the rise of consumer empowerment that has happened in virtually every other sector. In a fourth-party payment system (i.e., employer pays an insurance company which in turn pays the provider), the consumer is shielded from the true costs. When that shield is removed, one can expect them to have a higher expectation of communication with their providers.

CONCLUSION

The good news is that there is a tremendous competitive advantage that healthcare providers can realize if they choose to focus on improved communications for the 99+% of the time when their patients are not in their offices. Not only can this opportunity provide a competitive advantage, it is imperative in the new reimbursement models.

Fortunately, various studies referenced in this book demonstrate improved communication positively affects the goals of the Triple Aim. Improved communication leads to better outcomes, while empowered patients are also much more satisfied with the experience. At the same time, this has the effect of lowering healthcare costs. On top of all of that, clinicians express how much more fulfilled they are in their roles. The root of the word "doctor" is teacher. Doctors, as well as all other clinicians, went into the profession to improve the health of their patients and community; nothing is more fulfilling than guiding a patient back to greatly improved health.

Why communication is good for patients:

- Improved communications with providers has been correlated to improved health outcomes for patients.
- Patient/family satisfaction is improved and anxiety reduced with strong communication with the healthcare team.
- Better understanding of one's health situation leads to greater empowerment.

Why communication is good for providers:

- Improved communication capability increases operational efficiency.
- Providers express greater professional fulfillment.
- Enhanced communication leads to improved outcomes.
- Reduced network leakage of patients is enabled by top-of-mind status.
- Increased patient satisfaction leads to greater reimbursement (due to new Medicare requirements).

Why communication matters for the healthcare system as a whole:

- Payment reform is radically transforming the clinical delivery model. The change management that accompanies can be improved through strong communication.
- Improved health outcomes can slow or reverse overall costs.
- New communication requirements drive new and innovative technology solutions.

With 75% of healthcare spending going to chronic disease, communication is the best method for helping patients self-manage their conditions.

For more information, read the following case studies at the end of this book:

Case Study 1: Patient Engagement in a Patient-Centered Medical Home
Case Study 2: Innovation in the Context of Meaningful Use
Case Study 6: Kaiser Permanente—My Health Manager
Case Study 16: Online Therapy and Patient Engagement
Case Study 17: NeuCare Family Medicine, a Direct Primary Care Practice
Case Study 18: Total Access Physicians in Burlington, Kentucky
Case Study 19: Patient Engagement and Direct Practice
Case Study 20: Good Days Ahead, Online Cognitive Behavior Therapy
Case Study 21: Patient Portals Offer Opportunities for Patient-Provider Interaction at Children's Hospital of Philadelphia
Case Study 22: Veterans Health Administration, Department of Veterans Affairs—Patient Engagement and the Blue Button

REFERENCES

1. Southcentral Foundation's Nuka Model of Care: Rocks, birds and cars (Part 1 of 2). http://www.youtube.com/watch?v=tLnZ3_AccoU. Accessed August 25, 2012.

2. Zhou YY, Kanter MH, Jian J. Wang JJ, et al. Improved quality at Kaiser Permanente through e-mail between physicians and patients. *Health Aff.* 2010; 29(7):1370-1375. doi: 10.1377/hlthaff.2010.0048.

3. Kaiser Permanente. Power and convenience of Kaiser Permanente electronic health record drives member retention. http://thelundreport.org/resource/power_and_convenience_of_kaiser_permanente_electronic_health_record_drives_member_retenti_0 *The Lund Report.* Accessed August 25, 2012.

4. The Commonwealth Fund. Widespread Adoption of Information Technology in Primary Care Physician Offices in Denmark: A Case Study. *Issues in International Health Policy.* March, 2010.

5. Understanding growth charts. Seattle Mama Doc 101. http://www.youtube.com/watch?v=azxpCM75NOU, Accessed August 25, 2012.

6. Hemoglobin A1c: A candy shell. http://www.youtube.com/watch?v=-EN0Ktx7dvE. Accessed August 25, 2012.

7. Tylenol 2.0: The new infant acetaminophen. http://www.youtube.com/watch?v=IHobFuLFLvU. Accessed August 25, 2012.

8. Knee Replacement – Just the facts. http://www.youtube.com/watch?v=yjkkNI3XDCk. Accessed August 25, 2012.

9. Patient Protection and Affordable Care Action. Section 1301(a)(3).

10. Net Promoter Study by Qliance: While health insurance had the lowest average NPS (-5%), Qliance had 79% compared to Google and Apple, which were 53% and 72%, respectively.

11. Dave Chase. CalPERS $7.0 billion potential health insurance bunker buster. *Forbes; July 5, 2012.* http://www.forbes.com/sites/davechase/2012/07/05/calpers-7-0-health-insurance-billion-bunker-buster/. Accessed August 25, 2012. (Note: The patient populations had an average or above average percentage of patients with chronic conditions.)

12. McNickle M. 9 ways future EHRs need to support ACOs. *Healthcare IT News.* http://www.healthcareitnews.com/news/9-ways-future-ehrs-need-support-acos. Accessed August 25, 2012.

13. Epperly T. *Fractured.* New York City: Sterling & Ross Publishers; 2012.

14. Berwick DM. *Escape Fire: Lessons for the Future of Health Care.* The Commonwealth Fund; 2002.

15. National Ambulatory Medical Care Survey: 2008 Summary Tables # 1, 9, 13.

16. Zhou YY, Kanter MH, Jian J, Wang JJ, et al. Improved quality at Kaiser Permanente through e-mail between physicians and patients. *Health Aff.* 2010: 29(7):1370-1375 doi: 10.1377/hlthaff.2010.0048.

17. Dave Chase, Pioneer ACOs share lessons learned and challenges ahead. Forbes. June 12, 2012. http://www.forbes.com/sites/davechase/2012/06/12/pioneer-health-care-organizations-share-lessons-learned-and-challenges-ahead/. Accessed August 25, 2012.

18. Congressional Budget Office. Evidence on the costs and benefits of health information technology. May 20, 2008.

19. Epperly T. *Fractured.* New York City: Sterling & Ross Publishers; 2012.

20. Rau J. Patient grades to affect hospital medicare reimbursement. *New York Times*; November 11, 2011. http://www.nytimes.com/2011/11/08/health/patients-grades-to-affect-hospitals-medicare-reimbursements.html?pagewanted=all. Accessed August 25, 2012.

21. The emerging new workforce. http://www.littler.com/files/press/pdf/Emerging-New-Workforce-May-2009-Employer.pdf. Accessed August 25, 2012.

Chapter 7

Self-service Convenience: Benefits to Patients and Providers

By Jan Oldenburg, FHIMSS

INTRODUCTION

Our society puts a premium on convenience. We hustle through our lives, looking for the fastest way to complete our daily tasks. People shop, bank, and even donate money online because it is more convenient than the alternatives. Stored value cards and smartphone applications allow consumers to buy a cup of coffee without taking their credit cards out of their wallet or even carrying a credit card.

These trends are not going away. If anything, they are accelerating. A Pew Internet study posted in April 2012 noted that 61% of Americans use the Internet for online banking, making it as popular as using social media sites.[1] The same study noted that 71% of Americans use the Internet to shop, and earlier studies noted that 78% of Internet users agreed that online shopping was convenient.[2] The increasing prevalence of smartphones and sophisticated applications for them, as well as improvements in information technology overall, reinforce the expectation that services should be available at the click of a button, anytime, from anywhere.

These external factors affect consumer expectations of and attitudes toward healthcare. Consumers are anxious for healthcare to implement features that they already use elsewhere in their lives. A recent Accenture study, whose findings are reproduced in Figures 7-1 and 7-2, highlighted these desires, indicating that 73% of consumers want the ability to refill prescriptions online, 88% want to receive email reminders for preventive and follow-up care, and 72% want to be able to book, change, or cancel appointments online, and many of them want the same capabilities available on mobile devices as well.[3]

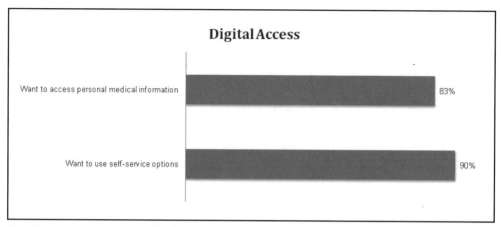

Figure 7-1. Patients' Preferences for Digital Access

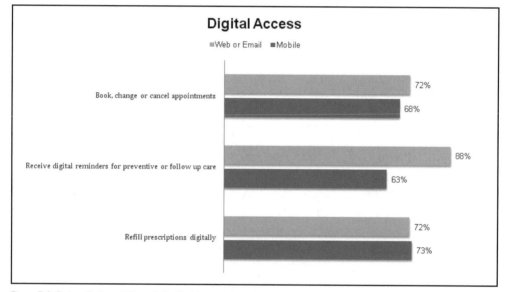

Figure 7-2. Patient-Preferred Channels for Healthcare-Related Tasks

Other industries have found that self-service features reduce the cost of their operations. For example, the cost to raise a dollar from a new donor using direct mail can be as much as $1.25 while telemarketing can cost $.63 to $.80 to raise a dollar; meanwhile, the cost to raise a dollar online can be as little as $.05.[4]

Online banking has a similar story to tell. Online banking, as well as self-service ATM machines, have allowed banks to save significant money on tellers and other staff while providing great customer satisfaction. "In 1996, the American Banking Association estimated that an online transaction costs $.01, an ATM transaction costs $.27, a telephone transaction costs $.54, and a branch transaction costs $1.07. This trajectory, in which the cost of a banking transaction completed online can be as low as 1% of an equivalent off-line transaction, has held true across time, changing the dynamics and the cost of banking."[5]

In retail, the story is similar—customers gravitate toward the convenience of online shopping, an environment in which the cost for every part of the operation is lower than in a traditional retail environment. As an example, a customer support call in an offline call center costs $15 to $20, while a self-service support request costs only $3 to $5 to service.[6]

In this chapter, we will explore the ways that self-service features offered in a healthcare context can follow this same trend—increasing consumers' satisfaction and meeting their desires for more convenience while creating efficiencies for the hospitals and clinics that implement them. Some of these features not only add convenience, but also support clinical goals.

The National Partnership for women and families interviewed eight providers who worked in practices using patient portals. According to one interviewee, "Portals are hugely efficient on the provider side for arranging things that don't need physician involvement—such as through the medication desk, referral desk, appointment desk, general desk. Instead of doctors having to deal with a lot of this individually, it's all taken care of for them, and they only have to deal with clinical questions."

Another provider explained that, "Those [providers] who complain the loudest are in the minority. Patients love it. We love it. It was totally worth the time and trouble of setting it up." Another noted, "Providing [patients with] more information, getting patients more involved, can result in the ability to spend more time and money on care." According to one physician, a patient portal is an "investment that pays off tremendously in saved work down the line."[7]

ONLINE APPOINTMENT SCHEDULING

Appointment scheduling online can take many forms. The most basic is an online appointment request, where a patient sends a request for appointment with preferred times and days. The online form then goes to the practice where employees actually book the appointment and notify the patient of the time. This approach may be slightly more convenient for a patient than a phone call, but because the patient does not have much control over the final day and time, it is not as satisfying as direct scheduling. This approach also does not affect practice efficiency because staff must still take direct action on each request.

Some urgent care clinics and emergency rooms publish wait times so that people can judge quickly where they will have the shortest wait time to be seen. Applications like iTriage and MedWaitTime provide this feature on mobile devices, and many hospitals and health systems offer their own version to gain competitive advantage. An example is Baptist Health's PineApp: "PineApp is an integral part of our ongoing commitment to improve the quality of care at a lower cost for our patients," said Petter Melau, Project Lead at Baptist Health. "Our patients use the app to decide which facility they should visit and what kind of a wait time they can expect from the emergency room and urgent care services of that facility when they arrive. In effect, we are load balancing our system while improving patient care."[8] Some applications also enable a patient to send a note informing the location that they are coming and describing their symptoms. This can support triage functions in the location.

The most flexibility for patients—and savings for the health system—occurs when clinics and hospitals initiate direct scheduling. In those settings, patients go online to choose a specific appointment that is right for them. If they have to reschedule an appointment, they can browse the list of available appointments to decide what will work best. There is evidence that patients who directly schedule their appointments online are more likely to show up for them. An internal Kaiser Permanente study in the Mid-Atlantic States region found that members using direct scheduling through kp.org to book primary care appointments are 33% less likely to no-show than members who book through the facility. In addition, 82% of cancellations through kp.org were made at least 24 hours prior to the scheduled appointment, allowing the facility to reuse the appointment slots.

Patients do not show up for appointments for many varied reasons, and reducing no-show rates can be a complex task. Online capabilities are only part of the solution, but in an environment in which many practices have no-show rates in the vicinity of 15% to 20%, with some running significantly higher than that, programs that reduce the number of no-shows can significantly benefit practice efficiency.[9]

Telephone calls and letters to remind patients of appointments have long been known to reduce no-show rates in clinic settings.[10] A Columbia University study that completed in February 2012 showed that text and email reminders can be as effective as letters and phone calls if the patient has chosen to receive reminders that way.[11]

Carolinas Healthcare is adding an alerting feature to its mHealth app to remind patients of upcoming appointments. Once a patient subscribes to the Carolinas web-based solution and sets up an appointment online, the application delivers an SMS (short messaging service) appointment alert. Since going live with the integrated solution, this healthcare organization has seen its appointment no-shows drop by 50%, providing a clear and demonstrable return on investment (ROI).[12]

These applications provide savings to the system while contributing to patient satisfaction. They also contribute to satisfaction on the part of providers and staff, as they translate to a workplace where visits can be better paced and expectations managed more effectively. As the American College of Physicians has noted, "Smooth and efficient appointment scheduling is critical to patient satisfaction, physician efficiency and staff morale."[13]

Physicians have expressed fears that patient direct scheduling will reduce their control over schedules, but those concerns can be eased by the way that appointments are released to the online scheduling system and by coordination with physician schedules.

Self-Service Appointments at MD Anderson Cancer Center

New cancer patients are anxious and generally contact several organizations trying to obtain a quick appointment. Because of its complexity, cancer requires acquisition of extensive medical record documentation before an appointment to reduce retesting and ensure correct diagnosis. At MD Anderson Cancer Center (MDACC), potential patients often began by filling out an appointment request form on the public website. Following completion of the online form, they get a call back from a staff member, confirming details such as insurance and other medical information prior to actually getting an appointment time. The wait time is excruciating to the patients, often prompting them to shop around to see if they can get an appointment sooner somewhere else.

Alicia Jansen, Vice President of Marketing at MD Anderson said, "We needed to streamline the process so we could meet our goal of offering appointments shortly after the request is made. By automating our systems to better engage our patients and improve clinical process flow, we can cut the time to schedule an appointment dramatically, and improve the patient experience significantly."

Tower Strategies assisted MDACC in creating a strategy that automated the web appointment process, resulting in patients receiving a tentative appointment by the end of the next business day. All of the intake processes are self-service, including monitoring medical record acquisition, insurance verifications, patient assessments, forms, permits, etc., to facilitate patients' experience even before they have a confirmed appointment and a medical record number. Future plans include a redesign of the current patient portal, mymdanderson.org, to provide a seamless, personalized experience from the patient's first touch point on the web through post-discharge and survivorship. Jansen noted, "Our goal is to offer a web experience for the patient that is integrated with other patient touch points and activities so it is seamless and personalized to the patient. Patients will have access to their own customized experience, at their convenience, anytime of the day or night."[14]

ONLINE PRESCRIPTION REFILLS

For providers, online prescription refills are part of a larger story about e-prescribing. E-prescribing brings a host of benefits to providers and patients alike, ranging from drug-drug interaction checking to formulary status verification at the point of prescription. Patient safety benefits also accrue from elimination of physician handwriting as an opportunity for errors. A 2012 Surescripts' study showed a consistent 10% increase in initial prescriptions actually picked up by patients when e-prescribing was used.[15]

Stage 1 of Meaningful Use requires that a practice use e-prescribing to send at least 40% of prescriptions. For physicians not participating in Meaningful Use Stage 1, the Medicare Improvements for Patients and Providers Act of 2008 (MIPPA) provides for a 2% payment incentive in 2011 for physicians using e-prescribing. A payment adjustment (penalty) was introduced for eligible professionals who have not implemented and employed a qualified e-prescribing system by June 30, 2011. The penalty is 1% of total allowed charges submitted throughout 2012 and a 1.5% payment cut for 2013 for the failure to e-prescribe.[16]

As you would expect, these requirements have driven a rapid rise in the use of e-prescribing. According to a 2011 Surescripts' study, the number of providers using e-prescribing increased by 67% in 2011, and the proportion of prescriptions refilled electronically by pharmacists increased by 75%.[17]

Prescription refills are generally an additional benefit when a practice implements a broader e-prescribing solution, but they are one of the benefits that most directly impacts patients. As noted, 72% of patients in the Accenture survey want to be able to request prescription refills online, and 73% want to have a mobile option for doing the same thing. A CVS CaremarkHealth IQ study in 2009 found that 28% of patients surveyed reported that they forget to refill their prescriptions on time, resulting in degradation of medication adherence; it seems likely that refill reminder systems would also benefit patients' medication adherence.[18]

To get the full benefit of online prescription refills, it is helpful both to include the capability to refill prescriptions online, as well as the ability to remind patients when a prescription needs to be refilled. Most community pharmacies offer the option for patients to refill prescriptions on their websites. Encouraging patients to use that route can positively impact physician workflows, as most prescriptions can be refilled directly with the pharmacy and do not require interactions with the practice at all. A 2008 study found that pharmacy staff made more than 150 million calls to physician practices a year to clarify prescription information—time that can be significantly reduced when electronic prescription and prescription refill practices are in place.[19] A Brown University study on the prescription renewals process before and after e-prescribing supports these findings. In this study, "The average prescriber time spent per day was cut in half, from 35 to 17 minutes, and the average staff time spent per day was cut in half from, 87 to 43 minutes."[20] Those minutes add up across a practice to real efficiencies in staff time and staff satisfaction.

---

ePrescribing's Impact

Linda Stolsky has been implementing electronic medical records (EMRs), health information exchanges (HIEs), and e-prescribing systems for 10 years, but before that, she was a hospital and clinic administrator. In an email exchange with the author, she noted, "As a practice administrator I was always on the phone with the pharmacy. Calling in scripts, formulary requests, correcting missing digits in demographic information—all resulted in lost productivity. In addition, there were errors. Leaving information on voicemail was never a good solution. The addition of electronic prescribing took this time drain away and immediately increased patient safety. Errors in handwriting were reduced. Prescriptions were formulary acceptable, based on the patient's insurance because mapping was done automatically. This means that the provider knows before prescribing whether the drug is accepted by the patient's insurance and can make the necessary changes at the point of the order.

"Allergies and adverse reactions are also taken into account and documented to address liability concerns. Malpractice insurance rates can sometimes be negotiated downward based on implementation of electronic medical records and e-prescribing. The value of having this information available at the point-of-care reduces adverse reactions. There are fewer deaths from anesthesia today because of the ability to coordinate past medical information and critical information at the point-of-care before a patient is given a drug that may be deadly."[21]

When prescription refill requests are integrated with an EHR, the requests can be presented as a task within the physician's daily workflow. The physician can check the patient's chart in the context of the request and make the decision on whether to refill the prescription based on clinical indications. The refill authorization can then be sent electronically to the pharmacy without intervention from staff and with no phone call to the pharmacy required.

This not only saves time for physicians and increases convenience for patients, but it also increases the speed with which patients can receive refills, making it more likely that patients' medication regimen will be uninterrupted. Increasing the speed with which patients receive refills also reduces repeat contacts between provider staff and patients about refill status.

The addition of text or email notices reminding patients that it is time to refill a prescription can also positively impact medication refill rates and patient medication adherence without adding staff time and costs. A 2012 Optum study found that text messaging patients to remind them to refill prescriptions significantly increased patients' refill rates.[22]

ONLINE VIEW AND PAY BILL

As noted, consumers are paying their bills online in record numbers, with more than half of U.S. consumers currently banking online. By providing the capability for patients to view and pay their bills online, health systems can add convenience for patients while reaping benefits such as reduced revenue cycle payment days, reduced staff to handle billing and follow-ups, and reduced use of collection services.

As consumers bear a greater portion of the cost of healthcare, it is increasingly important for health systems to consider these enhancements as a way to ensure that they are collecting appropriate revenues. The Centers for Medicare and Medicaid Services (CMS) estimates that over the last decade, consumers' out-of-pocket medical expenses rose from $146 billion to $249 billion and may reach $441 billion by 2016.[23] Even as health systems move to accountable care organizations (ACOs) and patient-centered medical home structures, it will be important to have convenient mechanisms to address payment for services that go beyond preventive and primary care. By 2011, more than 30% of hospitals in the U.S. had implemented an online bill payment function both for patient convenience and faster collections.[24]

We realize that the ability for patients to pay bills online is not a primary driver of patient satisfaction, especially if other aspects of their experience are cumbersome, frustrating, and manual. Electronic bill payment, however, has a place in building a suite of capabilities that add convenience for patients.

Practices can also support patients by providing them with the ability to view reports that show all of their healthcare expenditures with the system for the year.

This kind of self-service tool can simplify information collection for tax and funding purposes and can eliminate staff time needed to follow up on patient requests for such information.

Insurers are doing their part to add convenience to this part of the process as well, adding funding tools such as healthcare reimbursement accounts (HRAs) and healthcare savings accounts (HSAs), which may offer debit cards that enable a patient to pay for services directly at the point of care. These innovations can reduce the cost of collections by enabling up-front collection of co-payments or share-of-cost payments. In order to gain the full advantage of these services, health systems need to be as transparent as possible about the price in advance of service delivery.

COST TRANSPARENCY TOOLS

There is increasing pressure on providers of all kinds to provide for consumers price information in advance of the visit. A proposed House Bill in May of 2012,[25] now referred to Committee, would require states to put cost transparency requirements in place if they do not already have them:

"The Health Care Price Transparency Promotion Act directs the states to establish and maintain laws requiring disclosure of information on hospital charges. Additionally, the bill requires hospitals and health plans to make such information available to the public, and to provide individuals with information about estimated out-of-pocket costs for healthcare services. Thirty-four states already require hospitals to report information on hospital charges or payment rates and make that data available to the public; an additional seven states have voluntary efforts."[25]

Existing and proposed legislation has been put in place primarily to aid consumers in understanding likely out-of-pocket costs before they incur them, and both providers and insurers have been working to provide information that is accurate, timely, and easy-to-understand, while being clear about the potential for a range of costs if complications arise. At a minimum, healthcare systems are offering simple fee schedules online, so that patients can compare similar costs for similar services. Sites like www.VIMO.com[26] offer price comparisons by hospital and by procedure, and CMS[27] and the Department of Health and Human Services (HHS)[28] have begun offering quality comparisons online. More complex versions offered by insurers provide consumers with estimates that are informed by the patients' specific plan of benefits, the patients' progress against deductibles year-to-date, and whether their provider is in network. These personalized estimates enable patients to predict their out-of-pocket costs for services and may alter their decisions about whether to have procedures done or what providers to use.

These capabilities offer opportunities for healthcare systems as well. Providers who are clear about their prices and fees—especially those who are willing to discuss them with patients—win high marks for openness and honesty.[29] Thus, clarity about the cost of services can be a factor in retaining patients, building their trust and creating competitive advantage in the market.

There are also indications that patients who know what to expect when a bill comes and are not surprised by it are more likely to pay promptly—and are even significantly more likely to pay at the point of service. This increases up-front revenue collection and reduces the cost of collections. A study of six orthopedic practices noted that estimating patient liability actually doubled their collections on the day of service.[30] Other industry studies have noted that the likelihood of collecting fees is reduced by as much as 60% if the healthcare system bills for the service after the fact, rather than collecting at time of service.[31]

As consumers bear more of the cost of healthcare services, they are becoming increasingly savvy about the cost of care and their role in understanding and negotiating the cost of care. This creates an opportunity for physicians and patients alike to have a shared decision-making conversation about the cost and benefits of alternative treatments. Understanding a patient's values about the cost of care can be useful for the physician to be able to recommend treatments that provide the highest efficacy at a price that their patients can live with.

PRE-VISIT DATA COLLECTION

Many practices are implementing tools that enable patients to provide demographic and medical history information up front, eliminating the need for duplicative entry of the information on clipboards in the office each visit. Examples include No More Clipboard and Avado, both of which reduce duplicative tasks for patients during the visit and increase office efficiencies.

Some examples of the benefits include:
- Verification of information before the day of visit can streamline the registration process in the office and reduce the impact of patients who show up moments before the start of their visit. This can keep the practice on schedule.
- Information validated ahead of time can become a part of the chart and can enable the physician to take medical history into account more easily than working with information in several different forms.
- Pre-registration using kiosks can reduce the staff required to support the registration process.
- Pre-registration prior to hospital admissions have been estimated to save institutions both time and money.

GOING GREEN

Most of the emphasis on providing patients with access to their clinical data, such as laboratory results, immunizations, allergies, medications, past visit information, focuses on the clinical benefits of such access. And there are, of course, real clinical benefits from providing patients with access to their own clinical data in real or near real time as discussed in Chapter 5.

What is often overlooked, however, is that there are also efficiency gains for the practice in providing these capabilities online. The largest impact may be avoiding printing and mailing costs for notifying patients of laboratory results. A study at the Cleveland Clinic estimated the impact of providing test results online as saving $4,400 per provider per year, just based on the cost of stamps without factoring in the cost of paper, printing, envelopes, or labor.[32]

Many hospitals and healthcare providers are following their implementation of an EHR with efforts to go paperless both internally and externally. Kaiser Permanente exemplifies this approach by engaging in an effort to make any document that a patient or health plan member might otherwise receive by mail available online. They are engaged in an active campaign to inform their customers that the option is available and enlist them to help save the organization time and money on printing and mailing costs.

Kaiser Permanente's "Go Green" Initiative

At the end of 2010, Kaiser Permanente's expanding "go paperless" team reached a milestone. They calculated that their efforts to get members to "opt out" of paper documents in favor of seeing them online meant that they had saved the equivalent of 25% of the trees in New York's Central Park. The repository has now grown so much that by mid-2012, there are 180 million client and member documents in its document repository, equaling nearly half a billion pages of paper. The team calculates that they have now saved the equivalent of *two times* the trees in Central Park. The team's goal is to be able to deliver any and every document via a member's preferred channel including electronically. And members are signing up enthusiastically to go paperless.

The effort to date has focused on health plan documents and was largely limited to general content rather than the highly individualized content associated with specific member interactions with the health plan. Kaiser Permanente has been delivering Medicare Part D Explanation of Benefits (EOBs) electronically for several years now, despite concerns that the Medicare population would not be open to electronic documents. Surprisingly, more than 25% of the Medicare population has elected to receive their Part D EOBs online, which provides savings of approximately $535,000 for Kaiser Permanente annually.

Based on the success of the health plan's initial efforts in the go-paperless initiative, the team recently turned its attention to care delivery opportunities. The Southern California region has a high-cost, high-risk population of about 300,000 people with diabetes, coronary artery disease, chronic heart failure, and chronic kidney disease, referred to as their Cardiovascular Disease (CVD) population. This population needs regular laboratory screenings to ensure that problems are caught early, before they result in expensive hospitalizations. The screening rates also directly affect southern California's performance on Healthcare Effectiveness Data and Information Set (HEDIS) measures. The team explored whether they could send email to this population rather than paper mail reminders that they are due for a screening test without negatively affecting screening rates.

The pilot went live in April 2012. About 25,000 of the 300,000 patients in the CVD cohort receive screening reminders in any given month. Of those, 53% elected to receive their reminders online through My Health Manager on kp.org. A control group received no reminders at all. The remainder received paper letters reminding them to get care.

The results were impressive. Both groups that received reminders were significantly more likely to get the screening tests done than the control group. Twenty-six percent of the control group members came in for their recommended screenings. Slightly over 47% of the people who received paper-based reminders came in for screenings, and just over 43% of the people receiving email reminders came in for screenings. The cost difference, however, is dramatic between the two approaches to reminders: the cost per outcome is $3.71 to produce and mail each paper letter, while the cost per outcome to produce each email reminder is only $.33. Using email reminders, they are able to achieve essentially the same clinical results as with paper reminders at less than a one tenth of the cost. (Note that this is approximately equal to the proportionate drop in costs other industries report when they move transactions online.)

This pilot has been so effective that southern California is rapidly expanding the use of the outreach model and email reminders to other populations, and other Kaiser regions are exploring how they can adopt similar measures.

Rob Klusman, Director of Customer Service Technology Strategy and the interviewee for this vignette, had these words of advice for people who are leery of moving document delivery online. "Just remember, your customers have already been asking for these capabilities. You must have a good design so they can find the content they are looking for, but their reaction is more likely to be 'what took you so long' rather than 'why are you taking away my paper?'" Klusman also notes that the go-green initiative benefits from the already high usage of the kp.org portal; members are already in the habit of going to kp.org as a one-stop-shop for their clinical and health plan data.[33]

The potential impact in printing, mailing, and labor costs avoided if all paper currently sent to patients across the U.S. healthcare system were instead delivered electronically is enormous. The impact could mean reductions in the cost of healthcare overall, and it could also enable more money to be directed into clinical services rather than administration.

CONCLUSION

Self-service features offer a rare opportunity for the healthcare system to satisfy patients and customers by offering convenient ways to interact with the system as a whole—and in the process, enhance internal efficiency and save money: truly a win-win opportunity.

Why do self-service features matter for patients and consumers?

Consumers have come to expect to be able to perform many kinds of transactions online or on mobile devices 24x7, from banking to shopping. Offering similarly convenient transactions in the healthcare system eliminates a source of stress for patients and simplifies their lives. The added convenience can mean that patients are more likely to manage their health appropriately—make the appointment they need, refill a prescription on time to stay compliant with medications, and pay bills promptly.

How do patient self-service features matter for providers?

For providers, self-service features create opportunities to improve patient loyalty and satisfaction while creating a more efficient practice. The opportunities abound to create more efficient internal workflows—with accompanying reductions in direct costs and the opportunity to rebalance staff—as a result of providing patients with self-service transactions. Harvesting these opportunities can result in real savings, a more satisfied workforce, and competitive advantage in the market.

Why do self-service features matter for the healthcare system as a whole?

Self-service transactions create efficiencies for the healthcare system as a whole while increasing consumer and patient satisfaction. In an environment in which the cost of healthcare is unsupportable, it is critical that the system as a whole begin to employ efficiencies that other industries have shown to reduce costs.

For more information, read the following case studies at the end of this book:

Case Study 1: Patient Engagement in a Patient-Centered Medical Home

Case Study 4: From the Heart: The Intersection Between Personal Health and Technology

Case Study 6: Kaiser Permanente—My Health Manager

Case Study 12: Using *iTriage*® to Engage Healthcare Consumers

Case Study 16: Online Therapy and Patient Engagement

Case Study 20: Good Days Ahead, Online Cognitive Behavioral Therapy

Case Study 22: Veterans Health Administration, Department of Veterans Affairs—Patient Engagement and the Blue Button

REFERENCES

1. Zickuhr K, Smith A. Digital differences. Pew Internet. April 13, 2012. http://pewinternet.org/Reports/2012/Digital-differences/Main-Report/Internet-activities-Those-already-online-are-doing-more.aspx. Accessed September 23, 2012.

2. Horrigan J. OnLine shopping. Pew Internet. February 13, 2008. http://www.pewinternet.org/Reports/2008/Online-Shopping/01-Summary-of-Findings.aspx. Accessed October 1, 2012.

3. Accenture: Insight Driven Health. Is healthcare self-service online enough to satisfy patients? June, 2012. http://www.accenture.com/SiteCollectionDocuments/PDF/Accenture-Is-Healthcare-Self-Service-Online-Enough-to-Satisfy-Patients.pdf. Accessed September 23, 2012.

4. Network for Good. The young and the generous. http://www.fundraising123.org/files/The%20Young%20and%20The%20Generous%20A%20Network%20for%20Good%20Study.pdf. Accessed October 1, 2012.

5. Osho G. How technology is breaking traditional barriers in the banking industry: Evidence from Financial Management Perspective. *European Journal of Economics*. Finance and Administrative Sciences. 2008; 11. http://www.eurojournals.com/ejefas_11_02.pdf. Accessed October 1, 2012.

6. Strauss J. El-Ansary A, Frost R. Price. In *eMarketing*. New Jersey: Prentiss-Hall, 2003. www.nd.edu. Accessed October 1, 2012.

7. National Partnership for Women and Families. Using health IT to engage patients in their care: The role of providers in implementing online access. September 2011. http://www. nationalpartnership.org/site/DocServer/1639530218?d_opview=&docID=11261. Accessed September 23, 2012.

8. Mobile App. Emergency Room Waiting Time. Received Most Wired Award. September 20, 2011. http://www.24-7pressrelease.com/press-release/mobile-app-that-provides-emergency-room-waiting-time-to-patients-receives-most-wired-award-235995.php. Accessed September 23, 2012.

9. Lewis DP. Study's solution to no-shows: Overbooking. Amednews.com. February 4, 2008. http://www.ama-assn.org/amednews/2008/02/04/bil20204.html. Accessed September 30, 2012.

10. Bigby J, Giblin J, Elizabeth M, et al. Appointment reminders to reduce no-show rates: A stratified analysis of their cost-effectiveness. *JAMA*. 1983; 250(13):1742-1745.

11. Finkelstein SR, Liu N, Jani B, et al. Appointment reminder systems and patient preferences: Patient technology usage and familiarity with other service providers as predictive variables. http://www.google.com/url?sa=t&rct=j&q=&esrc=s&source=web&cd=1&ved=0CGEQFjAA& url=http%3A%2F%2Fwww.columbia.edu%2F~nl2320%2Fdoc%2FHealth%2520Informatics %2520reminder%2520systems.pdf&ei=6_AnULWbDuWoiQLj54GwDg&usg=AFQjCNHU6o yjxu8_m0DWaFAXXooEmgtLvw&sig2=1pxZROR5nxHrPinA2xFsXw. Accessed September 23, 2012.

12. Moore John; Moore, John III. mHealth Adoption for Patient Engagement, Chilmark mHealth Report. May 2012, page 16.

13. Center for Practice Improvement and Innovation. Designing the patient schedule. January 2010. www.acponline.org/running_practice/patient_care/pat_sched.pdf. Accessed September 30, 2012.

14. Sutton S. CEO Tower Strategies.

15. Surescripts Press Release. Study: E-Prescribing shown to improve outcomes and save U.S. healthcare system billions of dollars. February 1, 2012. http://www.surescripts.com/news-and-events/press-releases/2012/february/212_ePrescribing.aspx. Accessed September 30, 2012.

16. The Center for Improving Medication Management. A clinician's guide to e-Prescribing, 2011 update, Section 3. http://www.mgma.com/WorkArea/DownloadAsset.aspx?id=1248619. Accessed September 30, 2012.

17. Surescripts. The National Progress Report on e-Prescribing and interoperable healthcare 2011. http://www.surescripts.com/about-e-prescribing/progress-reports/national-progress-reports.aspx#downloads. Accessed September 30, 2012.

18. Redman R. CVS rolls out prescription refill reminder program. *Chain Drug Review*. August 13, 2009. http://www.chaindrugreview.com/newsbreaks-archives/2009-08-10/cvs-rolls-out-prescription-refill-reminder-program. Accessed September 30, 2012.

19. Electronic prescribing: Becoming mainstream practice. June 2008. http://www.thecimm.org/ PDF/eHI_CIMM_ePrescribing_Report_6-10-08.pdf. Accessed July 15, 2012.

20. Lapane KL, Whittemore K, Rupp MT, et al. Maximizing the effectiveness of e-Prescribing between physicians and community pharmacies. Brown University; 2007.

21. Linda Stolsky; email exchange with Jan Oldenburg.

22. Lee SH, Gu N, Kim B-H, Lim KS, et al. Impact of a Text Messaging Pilot Program on Patient Medication Adherence. *Clinical Therapeutics*. May 2012; 34(5):1160-1169. http://www. clinicaltherapeutics.com/article/S0149-2918%2812%2900265-2/abstract. Accessed September 30, 2012.

23. Grogan T, Albainy M. Engaging your patients in price discussions. *American Academy of Orthopaedic Surgeons (AAOS) News.* January 12, 2012. http://www.aaos.org/news/aaosnow/jan12/managing2.asp. Accessed September 30, 2012.

24. HIMSS Analytics. Essentials of the US Hospital Market, 7th Edition. www.himssanalytics.org/docs/18-NGRCM-12.pdf. *Accessed September 30, 2012.*

25. Dolan M. Burgess and Green Legislation Provides Patients Cost Information for Health Care. *About Health Transparency.* May 21, 2012, http://blogs.ipro.org/abouthealthtransparency/2012/05/21/burgess-and-green-legislation-provides-patients-cost-information-for-health-care/. Accessed September 30, 2012.

26. VIMO. http://www.vimo.com/hospital/cost.php. Accessed September 30, 2012.

27. Centers for Medicaid and Medicare Services (CMS). http://www.cms.gov/Medicare/Quality-Initiatives-Patient-Assessment-Instruments/HospitalQualityInits/HospitalCompare.html. Accessed September 20, 2012.

28. Health and Human Services (HHS). Hospital Compare. www.hospitalcompare.hhs.gov/. Accessed September 20, 2012.

29. Clarke, RL. Price Transparency: Building Community Trust. Financial Management Association (HFMA). *Front Health Serv. Management.* Spring, 2007; 23. http://www.ncbi.nlm.nih.gov/pubmed/17405387. Accessed September 20, 2012.

30. Grogan T, Albainy M. Engaging your patients in price discussions. *American Academy of Orthopaedic Surgeons (AAOS) News.* January 2012. Accessed September 20, 2012.

31. HIMSS. Re-engineering the Revenue Cycle for the Emerging Medical Consumer White Paper. 2008; page 14. www.himss.org/ASP/ContentRedirector.asp?ContentID=78491. *Accessed September 30, 2012.*

32. Nehring G. Patient Portals: A Jumpstart on Meeting Meaningful Use Requirements. *HIMSS News.* http://www.himss.org/asp/ContentRedirector.asp?ContentId=74627&type=HIMSSNewsItem. Accessed September 20, 2012.

Chapter 8

The Power of Mobile Devices and Patient Engagement

By Bianca K. Chung, MPH, with Brad Tritle, CIPP

INTRODUCTION

Ten years ago, cell phones, mobile email, handheld PCs, and pocket-sized video game consoles were considered a luxury, "nice to have" tech gear. Today, these items are necessities of daily life. Mobile devices permit the free flow of information and communication, capture life through video and still images, provide entertainment, and store libraries of digital books, notes, maps, and photos. Fewer than 1 billion people used a mobile phone[1] at the end of 2000. Now, there are over 6 billion active mobile subscriptions worldwide with over 30 billion software applications downloaded in 2011 alone.[2] Given the strong computing power and extensive storage found in tablets (e.g., iPad, Kindle, Samsung Galaxy among others), we can carry around a digital library, a lifetime of documents, and our entire family photo albums on a device the size of a magazine. Essentially, mobile devices have changed our lives.

As the mobile device market expands into tablets and wireless sensor technologies, more healthcare applications are unfolding with significant promise for empowering patients to actively manage their healthcare and/or facilitating the requirements of caregivers and providers. Healthcare is the fastest, most prominent growth area for mobile device applications.[3] Consider the following:

- Wireless home-based health applications and services are estimated to result in a $4 billion industry by 2013.[4]

- Over 80% of physicians own a mobile device, and 30% of physicians use a smartphone and/or tablet to treat patients as of 2012.[5]

- According to Research2Guidance 2010, 500 million smartphone users worldwide will be using a healthcare application by 2015.[6]

In this chapter, we will provide an overview of how mobile devices are currently used in healthcare applications by patients and providers and what the near-term potential is. This chapter will also identify the leading factors driving greater adoption of mobile devices and how the U.S. Food and Drug Administration (FDA) seeks to regulate their use in clinical applications.

A mobile device is a small, hand-held computing device weighing less than 2 pounds (0.91 kg), that has an operating system with various types of end-user software applications, apps, including health and fitness, entertainment, mobile banking, productivity tools that equip on-the-go end-users with mobile solutions to fit their needs. At this time, we are defining mobile devices as including the following categories of devices:

- Smartphones (e.g., iPhone, BlackBerry, Android®, Nokia N95, HTC Touch);
- Cell phones (without ability to access content from the Internet);
- Pagers (e.g., Vocera);
- e-book readers (e.g., Kindle);
- Portable navigation devices (e.g., global positioning system [GPS], Mapquest);
- Multi-media players (e.g., portable DVD player);
- Tablet PCs (e.g., iPad, Samsung Galaxy); and
- Handheld gaming devices (e.g., Nintendo 3DS, GameKing, Gizmodo, Play Station Vita).

Most mobile devices are equipped with camera, video, Wi-Fi, Bluetooth, and GPS capabilities that enable access to the Internet and wireless communication through a headset, microphone, and/or car speaker. In exchange for portability, mobile devices have limited processing power and battery life compared to alternatives.

Over three quarters of the world's population have a mobile phone. In the U.S., 88% of the population has a cellphone and about half of those are smartphones.[7] According to the latest NPD Quarterly Mobile PC Shipment and Forecast Report, overall mobile tablet shipments will grow from 347 million units in 2012 to over 809 million units by 2017.[8]

The Apple iPhone™ introduced apps in 2007, and their popularity skyrocketed from there. There are over 1 million apps available, and apps represent the fastest growing consumer segment in ecommerce.[9] In 2011, over 30 billion apps were downloaded to mobile devices.[10] Approximately 19% of smart phone users have downloaded at least one health app.[11] Most mobile devices have the ability to access content and interact with it through apps. This is shifting users away from desktop and laptop computers into mobile computing on handheld devices.

HOW ARE MOBILE DEVICES EMPOWERING PATIENTS TO ACTIVELY MANAGE THEIR HEALTH?

According to a Pew survey, 59% of U.S. adults look online for health information.[12] Considering that over 80% of the world's population owns a mobile phone, the potential for people using mobile devices to look up health information is quite significant. With greater adoption of people using health-related apps, the use of mobile devices to manage one's health and wellness is a compelling trend.[13] Consider the following:

- A survey performed in August to September of 2012 indicated that 31% of U.S. cell phone owners have used their phone to look for health or medical information online—an increase from 17% in 2010.[14]

- And 27% of Internet users, or 20% of adults, have tracked their weight, diet, exercise routine or some other health indicators or symptoms online.[15]

Leading categories of applications for patients using mobile devices are highlighted next.

Fitness, Nutrition and Wellness

As of April 2012, there were more than 13,600 iPhone health and fitness apps available for consumers.[16] A study presented to the American Heart Association found that obese and overweight patients who used an electronic logging system that was set up to provide personal feedback based on their goals were more likely to lose weight compared to control groups that used a paper log or an electronic log without feedback.[17] The participants who used an interactive electronic system were better able to maintain regular attendance at group sessions; meet daily calorie goals; meet dietary intake goals; achieve weekly exercise goals; and keep track of their eating and exercise patterns. Over time, all three groups regained some of their lost weight, but the group with the interactive electronic logs regained the smallest amount.[18]

Manage Chronic Disease

Glucometer apps collect blood glucose readings and caloric intake to help diabetic patients manage their pre-meal insulin doses and track their overall blood glucose levels. For example, glucose monitoring devices can connect to iPhones or the iPod Touch wirelessly via Bluetooth. The user inserts a test strip into the device, which collects the user's blood sample. The iPhone or iPod Touch confirms a correct reading and automatically syncs that data to their app, and stores up to 300 readings that are date and time stamped. According to a study by the University of Maryland School of Medicine, researchers found that patients using a mobile health app to track and monitor blood glucose levels, specifically the amount of hemoglobin A1c in their blood, decreased those levels by an average of 1.9% over a period of one year.[19] Charlene C. Quinn, PhD, RN, an Assistant Professor of Epidemiology and Public Health and the principal investigator of the study said, "The 1.9% decrease in A1c is significant. Previous randomized clinical trials have suggested that a 1% decrease in A1c will prevent complications of diabetes, including heart disease, stroke, blindness and kidney failure. Mobile asthma management tools include apps that help patients know where and what triggers asthma attacks in order to better prevent and treat asthma complications. The GPS location tool in the mobile devices tracks and logs where a person uses an inhaler. The asthma management apps identify the high trigger zones based on the data entered and include early warning software that can alert patients to potential asthma attacks.

Sensor Technology

The wireless sensor's main goal is to provide biofeedback in real time of the person's physiologic data, such as heart rate, blood glucose, and movement. The sensors are either attached to the body or monitors can be inserted into clothing, bracelets, badges, or even tattoos.[20] According to a study by GigaOM Pro, smart-skin tattoos can monitor blood glucose, drug therapy delivery or serve as a nicotine patch.[21] Other entrepreneurial start-ups are developing wearable patches that monitor kidney function, metabolism and blood glucose levels.[22] For example, a co-venture between the Gates Foundation and the University of California, San Diego's Jacobs School of Engineering, is developing a wireless pregnancy tattoo with an embedded sensor that transmits uterus contractions and heart rate of the mother and the fetus from the mobile phone to the cloud.[23]

Another example of sensor and wireless capabilities is the Smartheart device, which connects customized electrodes to an iPhone, BlackBerry, or Android device via Bluetooth and turns these mobile devices into EKG/ ECG recorders. It is the first FDA-cleared mobile EKG/ ECG device that records electrocardiogram (ECG) data. Consumers obtain their EKG/ECG rhythm strip in real time and can email this information to their physician from the device.[24]

Medication Reminders

Nearly 2 billion cases of poor medication adherence each year are avoidable.[25] Poor adherence to a medication program is responsible for one third to two thirds of medication-related hospital admissions.[26] Research studies have demonstrated the value of mobile medication reminder apps to help patients adhere to medication programs. A seven-month research study in 2011 looked at the efficacy of the Pill Phone application to improve medication adherence among hypertensive patients and found that it had positive results (n=50).[27] Patients had a high level of acceptance and sustained use of the Pill Phone application, and refill rates increased with the use of the application and decreased after the application was discontinued.[28] iRx, Care4Today Mobile Adherence, and Walgreens mobile app are all representative samples of apps focused on mobile medication reminders.

Text Reminders

Cell phones are ubiquitous around the world. Some 83% of American adults own cell phones and three quarters of them (73%) send and receive text messages. Cell phones are changing the way people can be connected, contacted, and involved in life events. According to Pew Research Center's Internet & American Life Project, 31% of people said they preferred texts to talking on the phone, while 53% said they preferred a voice call to a text message.[29] Another 14% said the contact method they prefer depends on the situation.[30] One indication of the power of text messaging comes from a study done by the Mobile Market Association that shows that it takes 90 minutes for the average person to respond to an email, while it takes 90 seconds for the average person to respond to a text message, as well as the fact that 70% of mobile searches result in action within an hour.[31] When a program called text4baby sent text messages to the 103,000 participants in the Pregnancy and New Mom protocol asking if they were planning to get a flu shot, more than 30,000 responded (39.8% already received a shot; 29.5% planned to get a shot; 31% did not plan to get a shot).[32] A separate study on the effectiveness of text4baby indicated that over 71% reported talking to their doctor after reading about a topic on a text4baby message.

The evidence that text messaging can help engage people, especially those with chronic diseases, in improving their health is so dramatic that Health and Human Services (HHS) Secretary Kathleen Sebelius established a task force that brought together public health experts from across HHS to ask how they could use mobile technologies more effectively. "We need to do a better job helping people manage their chronic diseases, so they avoid the worst complications."[33] Her group is focused on smoking cessation programs. According to Secretary Sebelius, the group found there are people who will not pick up a newspaper, will not turn on the radio, and may not read the flyer that their doctor gives them—but they will check their text messages.[34] Other programs aimed at awareness and health reminders, such as text4baby, which offers informational services sent via short messaging service (SMS) about prenatal care, nutrition, immunization, baby food recalls, etc.[35]

The Joint Commission, however, is taking an opposite stance, stating that physicians and other healthcare professionals should not use text messages as a way to share patient health information, especially in situations where physicians text orders to the hospital or other healthcare environment.[36] This is due in part to the lack of security controls within SMS at this time.

Health Video Games

According to iConecto, a research group tracking health video games, the Health eGames industry was already a $6.7 billion industry in 2007. Gamers were spending their own money for brain fitness exercises, condition management, and healthy behavior games.[37] There are apps that engage family members in planning healthy meals, enabling them to create customized profiles based on allergies, nutritional requirements, preferences and using the information to plan nutritional, inexpensive meals. There are apps for children that teach them to make good eating and nutritional choices through feeding a virtual pet. Children can see their virtual pet's daily nutritional requirements and determine whether their food choices meet their pet's personal food requirements. According to Debra Lieberman, Director of the Health Games Research program at the University of California at Santa Barbara, "Digital games are interactive and experiential, and so they can engage people in powerful ways to enhance learning and health behavior change." Emerging studies are showing this to be true. One study showed that patients who played Re-Mission, a video game developed by HopeLab in Redwood City, California, took their medications more consistently and absorbed cancer-related knowledge faster than those who did not play.[38]

Find a Physician, Schedule Appointments, and Use Appointment Reminders

Apps act as a coordination tool to connect patients to the right provider. For example, Kaiser Permanente launched an iPhone app in 2011 that located the closest Kaiser healthcare system or urgent care facility for its members. The My Health Manager app allows Kaiser members to access Kaiser's mobile website, making it easier to refill prescriptions, view previous visits and lab results, and schedule or cancel an appointment. Northern California members can use the KP Preventative Care app that acts as a reminder system for appointments and preventative screenings, as well as sends health tips and enables members to make appointments.

Blue Cross Blue Shield of Arizona (BCBSAZ) has recently launched AZBlue, its first mobile app with similar functionality. According to Elizabeth Messina, Senior Vice President and Chief Information Officer for BCBSAZ, "The new AZBlue app represents our commitment to serving our customers by providing them with the tools and services that enable easy access to information and improve their quality of life."[39]

The iTriage app offers a "symptom to provider" pathway where patients can self-select their symptoms from a list provided by the app and be able to find a nearby medical facility to seek further treatment. iTriage has taken this a step further with emergency department wait times and mobile appointment setting for patients and their medical providers.[40]

HOW ARE PROVIDERS USING MOBILE DEVICES TO ENGAGE PATIENTS IN MANAGING THEIR HEALTH?

Mobile devices are changing the way providers deliver care, and providers are positively responding to having greater flexibility in accessing content using a mobile platform in their job functions. Apps that run on these mobile devices lead the way in encouraging this transformation. Discussed next are the most desirable capabilities that mobile devices currently offer to providers.

Real-time Physiologic Data and Access to Images

Apps can retrieve physiologic data from the patient's bedside monitor and display it on a mobile platform within four seconds or less. These data include electrocardiogram or electroencephalogram data, cardiac monitors, pulse rates, arterial oxygen saturation, and other bedside devices. Some of the apps include alarms and alerts that are activated when a patient reaches a critical threshold. In the case of labor and delivery, apps can deliver critical real-time data to providers, enabling them to monitor their patient remotely and to indicate to the expectant parents when it is time to come to the hospital. For example, the Airstrip OB app allows doctors to check fetal heart tracings and contraction patterns originating from the hospitals' labor and delivery room in real time on their mobile phone.[41]

Monitoring high-risk patients returning from the hospital has demonstrated positive clinical outcomes and reductions in healthcare expenses. For example, the New England Healthcare Institute's (NEHI) updated 2009 report on following heart failure patients post-acute with remote monitoring equipment and software installed in the patient's home found a 60% reduction in hospital readmissions compared to standard care, and a 50% reduction in hospital readmissions compared to disease management programs without remote monitoring.[42] NEHI believes remote home monitoring of high-risk heart failure patients has the potential to prevent between 460,000 and 627,000 heart failure-related hospital readmissions each year.[43] Based on this reduction in hospital readmissions, NEHI estimates an annual national cost savings of up to $6.4 billion dollars.[44]

Mobile devices can access medical images, such as magnetic resonance imaging (MRI) scans, mammography data, x-ray, and computed tomography (CT) data, and process the image by featuring the capability for the provider to pan, tilt, zoom, measure, and add contrast to the image. Apps can connect to Digital Imaging and Communications in Medicine (DICOM) medical image servers and retrieve these data for the provider to assess and/or share information among the care team members and the patient. A recent study by the Radiological Society of North America found a majority of radiologists were able to correctly diagnose appendicitis by accessing Picture Archiving and Communication Systems (PACS) using an iPhone app OsiriX, a medical imaging analysis smartphone application that allows physicians to access records. The study found that in 125 total viewings of pelvic and abdominal x-rays (25 cases examined by 5 radiologists each), an accurate diagnosis was made 124 times.[45]

Diagnosis and Treatment

Mobile devices have become a diagnostic tool. Apps can turn a mobile device into a stethoscope via a transducer that amplifies the body's sound. Some apps state this type of functionality is for educational purposes, while others are saving lives. For example, StethoCloud, a custom built stethoscope and mobile phone app system, analyzes a person's breathing to determine if he or she has pneumonia. The specially designed microphone plugs into the smartphone's audio jack and captures the sounds of the person breathing. The app uploads the recording onto cloud servers, analyzes the breathing patterns, makes a diagnosis according to the standards of the World Health Organization, and then presents the user with the appropriate treatment plan.[46] StethoCloud costs about $20, whereas a regular digital stethoscope runs over $600. This can have significant impact in developing countries where diagnosing pneumonia in early stages can save lives.

Provide Health Education to Patients

The combination of ease of use, size, long-lasting battery power, and relatively low cost was the right mix for providers to see mobile devices as the next best thing to a paper chart. For example, Children's Hospital of St. Louis, Missouri, currently uses iPads to show patients how physicians prepare for surgery, play games with patients, and educate them about their procedures or conditions. This is aimed to ease the stress children experience prior to surgery.[47] Providers have found their mobile platforms also help patients understand their prognosis better, which helps increase their adherence to the post-hospital treatment plan.

High Growth of Mobile Capabilities

While the functional capabilities of apps are expanding, the mobile devices that operate them have seen incredible growth and adoption across every major demographic group—men and women, younger and middle-aged adults, urban and rural residents, the wealthy and the less well-off. The growth factors behind the widespread adoption of mobile devices are driven by the following leading trends:

- As costs decrease, mobile devices have more relative affordability.
- Mobile data traffic and the cellular coverage to manage it are increasing.
- Accessing content from mobile devices will be significantly faster with improved infrastructure.

As Cost Decreases, the Relative Affordability of Mobile Devices Enables More People to Adopt Them

Income plays a major role in mobile device adoption. Those with higher incomes are more likely to own a smartphone and/or a tablet PC. According to the NPD Group, the average price of a smartphone was $135 in 2011.[48] A recent study from the market research firm Nielsen found that among 18-24 year olds (who represent 17.2% of total smartphone users in the U.S.), over half of the respondents who make less than $15,000 a year said they own a smartphone. Those in the 55-64 age category making over $100,000 a year are almost as likely to have a smartphone as those in the 35-44 age bracket making $35,000-$75,000 per year.[49]

Mobile Data Traffic Is Growing Exponentially and Cellular Coverage Is Increasing

Smartphones, mobile PCs, and tablets with cellular connectivity generate five-to ten times more traffic than any other communication and/or computing device.[50] These mobile devices comprise around 50% of data traffic volume, and, as it is a fast-growing segment purportedly, will represent the vast majority by 2017.[51] Mobile data traffic is expected to grow with a compound annual growth rate (CAGR) of around 60% (2011-2017), driven mainly by video.[52] According to the study by Ericsson, the total subscriptions of data-heavy devices (includes smartphones, mobile PCs, and tablets with cellular connectivity), will grow from around 850 million by the end of 2011 to around 3.8 billion in 2017. As a result, coverage of the world's mobile networks is expanding.

Balancing the Delicate Act of Providing Security, Innovation, and Regulation

The demand for instantaneous mobile computing to now access healthcare has led the FDA to become concerned with certain apps that run on these mobile devices.

The FDA believes that this subset of mobile apps poses the same or similar potential risk to the public health as currently regulated devices if they fail to function as intended. Specifically, the categories of regulated medical devices referred to are those that are used for the purpose of displaying, storing, analyzing, or transmitting patient-specific medical device data.[53] The FDA's intention is both to provide for patient safety in this emerging smartphone for healthcare segment, while enabling growth and innovation as well. Tables 8-1 and 8-2 outline the apps that the FDA considers may be within their scope in requiring regulatory scrutiny and those that will be exempt, according to draft guidelines released in mid-2011.

Table 8-1. Mobile Apps That May Potentially Be Exempt from FDA Regulation (Source: Draft Guidance for Industry and Food and Drug Administration Staff Mobile Medical Applications. Document issued on July 21, 2011.)

- Mobile apps that do not contain any patient-specific information and function as electronic "copies" of medical textbooks, teaching aids, or reference materials (e.g., the electronic *Physician's Desk Reference*).

- Mobile apps that are solely used to log, record, track, evaluate, or make decisions or suggestions related to developing or maintaining general health and wellness. These apps are not intended for curing, treating, seeking treatment for mitigating, or diagnosing a specific disease, disorder, patient state, or any specific, identifiable health condition. Examples include tracking logs, appointment reminders, dietary suggestions based on a calorie counter, posture suggestions, exercise suggestions, or similar decision tools that generally relate to a healthy lifestyle and wellness.

- Mobile apps that only automate general office operations such as billing, inventory, appointments, or insurance transactions. Examples include apps that determine billing codes like ICD-9.

- Mobile apps that perform the functionality of an electronic health record system or personal health record system.

Table 8-2. Mobile Apps That Are Included in FDA Regulation Draft Guidelines (Source: Draft Guidance for Industry and Food and Drug Administration Staff Mobile Medical Applications. Document issued on July 21, 2011.)

- Mobile apps that are an extension of one or more medical device(s) by connecting to such device(s) for purposes of controlling the device(s) or displaying, storing, analyzing, or transmitting patient-specific medical device data. Examples include apps that display patient-specific data from medical devices, such as bedside monitors, stored EEG waveforms, and medical images. Additionally, apps that remotely control medical devices, such as blood pressure cuffs and insulin delivery on an insulin pump by transmitting control signals to the device from the mobile platform.

- Mobile apps that transform the mobile platform into a medical device by using attachments, display screens, or sensors or by including functionalities similar to those of currently regulated medical devices. Examples include apps that function as a stethoscope, a blood glucose strip reader or a glucometer, electrocardiograph (ECG), or a mobile app that uses the built-in accelerometer on a mobile platform to collect motion information for monitoring sleep apnea.

- Mobile apps that allow the user to input patient-specific information and—using formulae or processing algorithms—output a patient-specific result, diagnosis, or treatment recommendation to be used in clinical practice or to assist in making clinical decisions. This includes apps that analyze or interpret data (electronically collected or manually entered) from another medical device. Examples include mobile apps that provide a questionnaire for collecting patient-specific lab results and compute the prognosis of a particular condition or disease, perform calculations that result in an index or score, calculate dosage for a specific medication or radiation treatment, or provide recommendations that aid a clinician in making a diagnosis or selecting a specific treatment for a patient.[54]

CONCLUSION

Why do mobile devices matter for patients?

They are the conduit for health apps that give tools to patients to actively manage their health. It helps patients adhere to programs that enable them to attain their health goals as long as they engage in tracking their progress.

How do mobile devices matter for providers?

The utility of accessing real-time and/or historical relevant patient data from a mobile device creates a data-enriched, tele-presence experience.

Why do mobile devices matter for payers?

The pairing of wireless healthcare solutions with mobile devices has demonstrated significant reductions in healthcare costs and improvement in clinical outcomes for diabetic patients and patients afflicted with cardiopulmonary heart disease and asthma, as well as reduced readmissions for heart attack survivors.[55]

Why do mobile devices matter for healthcare overall?

The use of mobile devices by providers makes remote patient monitoring become relevant in the healthcare delivery system. In addition, mobile devices provide opportunities to expand patient engagement by providing education, reminders, trackers, and alerts that are contextually relevant and may change patients' ability to care for themselves.

For more information, read the following case studies at the end of this book:

Case Study 5: Social Media, Mobile and Gamification for Diabetes Care

Case Study 10: The E Is for Engagement

Case Study 12: Using *iTriage®* to Engage Healthcare Consumers

Case Study 14: Moving the Point of Care into Patients' Daily Lives via Text Reminders

Case Study 20: Good Days Ahead, Online Cognitive Behavioral Therapy

Case Study 23: Prescribing Apps for Weight Loss

REFERENCES

1. Kelly T, Minges M. Information and communications for development 2012. World Bank. IC4D 2012.

2. Ibid.

3. Asmundson P. Open mobile: the growth era accelerates. Deloitte Open Mobile Survey. 2012.

4. Park Associates. Wireless Home Healthcare to Be $4 Billion Industry by 2013. August 5, 2009.

5. Boone E. 5 ways mobile tech can improve your health. June 1, 2012.

6. Research2Guidance.

7. Kelly T, Minges M. Information and communications for development 2012. World Bank. IC4D 2012.

8. The NPD Group. As Smartphone Prices Fall, Retailers Are Leaving Money on the Table. Port Washington, NY; November 14, 2011. https://www.npd.com/wps/portal/npd/us/news/pressreleases/pr_111114a.

9. App. NPD report. Apps as fastest growing ecommerce segment.

10. Smith A. 46% of American adults are smartphone owners. Pew Internet Project. March 1, 2012.

11. Fox S, Dugan M. Mobile Health 2012. Pew Internet Project. November 8, 2012. Accessed November 11, 2012.

12. Fox S. Health Topics. Pew Internet and American Life Project. February 1, 2011.

13. World Bank. Information and Communications for Development 2012. July 2012.

14. Fox S, Dugan M. Mobile Health 2012.

15. Fox S. Health Topics. Pew Internet and American Life Project. February 1, 2011.

16. Dolan B. An Analysis of Consumer Health Apps for Apple's iPhone. 2012. July 11, 2012.

17. Styn M, Conroy M, Ye L, et al. American Heart Association Meeting Report. Abstract 026. March 15, 2012.

18. Ibid.

19. Quinn CC, et al. Cluster-randomized trial of a mobile phone personalized behavioral intervention for blood glucose control. *Diabetes Care.* 2011; 34(9):1934–1942.

20. Giga OM. Pro Mobile Health Report. July 2012.

21. Ibid.

22. Sano Intelligence. http://sanointelligence.com/.

23. Ibid.

24. Dolan B. FDA clears Smartheart mobile ECG device. *Mobihealthnews.com.* Accessed April 17, 2012.

25. 3 IMS Health; 2010. National Prescription Audit PLUS. http://www.imshealth.com/deployedfi les/imshealth/Global/Content/StaticFile/Top_Line_Data/2010_ Top_Therapeutic_Classes_by_ RX.pdf. Accessed November, 2011.

26. 3 IMS Health; 2010. National Prescription Audit PLUS. Retrieved from http://www.imshealth. com/deployedfiles/imshealth/Global/Content/StaticFile/Top_Line_Data/2010_ Top_ Therapeutic_Classes_by_RX.pdf. Accessed November 2011.

27. *PRNewswire*; 2011. Research study looks at benefits of 3G wireless technology for hypertensive patients in underserved urban communities. February 9, 2011. http://www. fiercehealth care.com/press-releases/research-study-looks-benefits-3g-wireless-technology-hypertensive-patients-. Accessed December 28, 2011.

28. Ibid.

29. Smith A. Senior research specialist. Americans and text messaging. Pew Research Center. September 2011. http://pewInternet.org/Reports/2011/Cell-Phone-Texting-2011.

30. Ibid.

31. Mobile Marketing Association. 2012 Mobile Fact Sheet. http://www.eztexting.com/sms-marketing-resources/2012-mobile-fact-sheet/. Accessed September 30, 2012.

32. Johns Hopkins University School of Nursing. Text4baby helps increase mom's flu season awareness & conversation. December 2, 2011. http://nursing.jhu.edu/news-events/news/ archives/2011/text4baby.html. Accessed November 11, 2012.

33. HHS Secretary Kathleen Sebelius. Digital Health for Digital Development Meeting. September 19, 2010. New York, NY.

34. Ibid.

35. http://text4baby.org/index.php/about/message-content.

36. http://www.jointcommission.org/standards_information/jcfaqdetails.aspx?StandardsFaqId=40 1&ProgramId=1.

37. Goldstein D, Groen P, Jaisimha J. Mobile Health Apps Market Analysis–Highlights. *iConecto*.

38. Chiang OJ. Apps and videogames to keep you healthy. *Forbes*. January 20, 2010. http://www.forbes.com/2010/01/20/iphone-apps-videogames-technology-breakthroughs-health.html. Accessed January 20, 2010.

39. Blue Cross Blue Shield of Arizona Releases New Mobile App–FierceHealthPayer. http://www.fiercehealthpayer.com/press-releases/blue-cross-blue-shield-arizona-releases-new-mobile-app#ixzz24swOF8Bx.

40. Hudson P. Case Study: Using iTriage to Engage Health Care Consumers.

41. http://www.airstriptech.com/.

42. New England health Institute. Remote Physiological Monitoring Update 2009.

43. Ibid.

44. Ibid.

45. Choudhri A. Johns Hopkins University School of Medicine in Baltimore. Preliminary findings of diagnosing appendicitis by smartphone imaging applications. 2009.

46. Hill DJ. StethoCloud–The $20 stethoscope attachment for smartphones to diagnose pneumonia. http://singularityhub.com/2012/08/10/stethocloud-the-20-stethoscope-attachment-for-smartphones-to-diagnose-pneumonia. Accessed August 10, 2012.

47. Dolan B. iPad vs. the Tablets in Healthcare. Accessed July 11, 2012.

48. The NPD Group. As smartphone prices fall, retailers are leaving money on the table. Port Washington, New York.https://www.npd.com/wps/portal/npd/us/news/pressreleases/pr_111114a. Accessed November 14, 2011.

49. Nielson. Smartphone growth by age and income. http://blog.nielsen.com/nielsenwire/online_mobile/survey-new-u-s-smartphone-growth-by-age-and-income. Accessed February 20, 2012.

50. Erikson. Traffic and Market Report. June 2012.

51. Ibid.

52. Ibid.

53. http://www.fda.gov/MedicalDevices/DeviceRegulationandGuidance/default.htm, Two sections: Overview of Medical Device Regulation. How to Market Your Device.

54. Draft Guidance for Industry and Food and Drug Administration Staff Mobile Medical Applications. Document issued July 21, 2011.

55. Khan JS. Participatory Health: Online and Mobile Tools Help Chronically Ill Manage Their Care. California Healthcare Foundation. September 2009.

Chapter 9

Social Media: An Engine for Transformation?

By Brad Tritle, CIPP

INTRODUCTION

What is social media? According to Gartner's Anthony Bradley, social media is "an online environment established for the purpose of mass collaboration."[1] The functionality developed for these online environments is also now being used by more closed, private, and secure systems to encourage greater collaboration.

Social media is a relatively new set of tools and functionality for healthcare, however, and one that is often not understood in its full breadth. Over 1 billion people are on social media platforms around the world via mobile phones, laptops and desktop computers, making it a natural location to both "find" and "engage" with existing and new patients.[2] But social media is much more than the household name for platforms such as Facebook, Twitter, and YouTube. Social media is functionality that already is being applied to a variety of traditional healthcare functions requiring communication, such as coordination of care, patient education, provider education, patient monitoring, and more. These uses will only grow in the future.

Healthcare providers, though individually participating personally in social media platforms at a higher than average rate, are not participating professionally at the same rates.[3] Ed Bennett, who tracks hospital participation in social media, estimates that only 21% of U.S. hospitals, including systems and integrated delivery networks (IDNs), are officially active on any social media platform.[4] Individual physicians, by comparison, participate at higher rates—with surveys reporting personal participation anywhere from 42% to nearly 90%, and professional participation at up to 65%.[5]

Social media benefits patients by meeting them "where they are" (e.g., 1 billion people on Facebook), allowing them to connect with others with the same condition to obtain support or education.[6] In social media, individuals can control how actively they participate, as well as how much personal information they reveal. Patients can also choose to participate in sites that engage them in health or wellness activities via "gamification" techniques designed to make the engagement experience pleasurable.[7, 8]

For providers, the benefits of social media use can include more adherent patients, increased patient acquisition, more efficient patient education (time saved), increased awareness of recent industry developments, and rapid establishment of a secondary communications channel for use during disasters.[9, 10, 11, 12, 13, 14]

In this chapter, we will provide an overview of social media, review how it is used by patients, explore how it is used by providers to engage patients, and what some leading-edge examples point to as the potential future uses by providers for patient engagement.

Social media online environments most commonly fall into the following categories:

- Social networking sites (e.g., Facebook, LinkedIn)
- Blogs (web-logs; online journaling in reverse chronological order)
- Content communities (e.g., YouTube)
- Collaborative projects (e.g., wikis)
- Crowdsourcing (asking the online community to solve a problem or develop or vote on new ideas/products)
- Microblogs (e.g., Twitter)
- Location-based apps (e.g., FourSquare)
- Rating sites (e.g., Yelp)

Please note, however, that some categories often are functionally incorporated into other categories (e.g., Facebook incorporating location-based functionality), some environments cross categories (e.g., Instagram is a content community that incorporates social networking; LinkedIn allows for the sharing of content, such as presentations), and users may choose to use a particular category in a unique way (e.g., using social networking platforms to perform crowdsourcing).

HOW ARE PATIENTS USING SOCIAL MEDIA?

Is social media influencing patients' health and wellness decisions? A recent survey by PricewaterhouseCoopers' (PwC's) Health Research Institute (HRI) indicated consumers are influenced by social media as they make the following decisions:

- 45% regarding whether to seek a second opinion;
- 42% regarding how they deal with a chronic disease;
- 42% regarding their approach to diet, exercise, and stress management;
- 41% in their choice of a hospital, medical facility;
- 41% in their choice of a specific doctor;

- 34% regarding taking a certain medication;
- 33% regarding undergoing a certain procedure or test; and
- 32% regarding choice of insurance.[15]

The following statistics from the Pew Research Center's Internet and American Life Project are also telling:

- 25% of all adults have read someone else's experience or commentary about health issues in a blog, news group, or website;
- 19% have watched an online video about health issues;
- 18% have consulted online reviews of drugs or medical treatments;
- 12% have consulted online reviews of doctors or other providers; and
- 11% have consulted online reviews of hospitals or other facilities.[16]

Knowing that patients are making health-affecting decisions based on these activities should be of great interest to providers. Let us delve a little deeper into patient activities to understand how patients' expectations of and interactions with the healthcare system are changing as a result of their social media engagement.

Education (Patients)

As the statistics indicate, consumers seek information through social media—they are seeking to become more educated. This can range from communicating with friends on general social networking sites such as Facebook, to watching health-related videos on YouTube (two surveys have indicated Facebook and YouTube are the most popular platforms used by consumers to search for health information), to engaging in a disease-specific community in order to learn more about treatments, side effects, or medications, or to gain access to disease-specific publications.[17]

Patients may also seek to stay up to date on developments around their condition by following individuals or organizations on Twitter, or by following specific "hashtags." Those posting on Twitter can index their tweets by including hashtags such as the following, to enable individuals to see tweets from anyone posting on a given subject:

- #BCSM (breast cancer)
- #Psoriasis
- #Diabetes
- #ColonCancer
- #Parkinsons
- #Rheumatoid Arthritis
- #Dementia

The Healthcare Hashtag Project, sponsored by Symplur, seeks to keep track of healthcare-related hashtags in use on Twitter, including those related to diseases, events, and other topics such as "Healthcare Social Media" (#hcsm), ePatients (#epatients), or a specific "tweetchat" on a given subject, such as #solvePT— which seeks to facilitate community discussions and problem solving around the subject of physical therapy. [18] A tweetchat is a discussion scheduled on Twitter by someone emerging as a host/moderator, often lasting an hour and held once a week, wherein the host may choose to focus the discussion on a shortlist of topics within the hashtag subject. The Healthcare Hashtag Project maintains a calendar of tweetchats that may be of interest to patients and providers at http://www. symplur.com/health care-hashtags/tweet-chats/.

Education by Blog

Patients are also reading and writing blogs, to share their own experiences and to learn from others' experiences. Examples include:

- Trisha Torrey was independently diagnosed with cancer by two laboratories, but discovered due to her own persistence, personal research, and consultation with other providers that she indeed did NOT have cancer—just weeks short of starting chemotherapy. She began an online forum called AdvoConnection to connect patients with patient advocates, and also moderates a patient empowerment forum on About.com. Her blog at TrishaTorrey.com is entitled "Every Patient's Advocate: A Blog about patient empowerment, advocacy, safety, consumerism, and tools to navigate the dysfunction of American healthcare."[19, 20]

- Kerri Malone blogs at sixuntilme.com and was one of the first to blog about her experiences with type 1 diabetes. Her website's subtitle is "Diabetes doesn't define me, but it helps explain me."[21]

- The Society for Participatory Medicine (#s4pm), a non-profit organization dedicated to furthering the concept of participatory medicine by and among patients, caregivers, and their medical teams, maintains a blog at e-patients.net, led by the Society's founder Dave deBronkart, famously known as "e-Patient Dave." It defines "participatory medicine" as "a cooperative model of healthcare that encourages and expects active involvement by all connected parties (patients, caregivers, healthcare professionals, etc.) as integral to the full continuum of care. The "participatory" concept may also be applied to fitness, nutrition, mental health, end-of-life care, and all issues broadly related to an individual's health."[22] Those that choose to become a member of the society can also participate in an ongoing listserv/forum-based discussion. Members include patients/consumers, physicians, nurses, patient advocates, government employees, consultants, vendors, and others. The Society also publishes the peer-reviewed, open-access *Journal of Participatory Medicine.*[23]

Patients who blog about their experiences hope that by sharing they may help others. They are examples of the "self-actualized" state of engagement described at the beginning of this book.

Many patients also follow physician and hospital blogs, which vary greatly in the topics covered. For example, John Halamka, MD, CIO of the Beth Israel Deaconess Medical Center and Harvard Medical School in Boston, started writing about his wife's journey with cancer on his blog in order to help other patients learn from their path.[24] We address more physician blogs later in this chapter.

Communicate/Community (Patients)

Thirty-two percent of consumers access social media to view family members'/ friends' health experiences, while 29% view other patients' experiences with a disease.[25] Sometimes this may occur on a large social network like Facebook, when a consumer interacts with friends and family via posts to their own or others' walls. On Facebook, advocacy organizations, healthcare providers or others may devote a site to promoting awareness, raising funds, and communication, such as Facebook pages of The Breast Cancer Site and the Mayo Clinic Guide to a Healthy Pregnancy. Facebook also may create non-sponsored community pages on its own, usually populating the page with general disease information, with options for an organization or individual to create a page on that topic, such as shingles.[26] Once an individual "likes" a Facebook page, they will receive updates on their wall from the "liked" page.

There are also large social networking platforms such as PatientsLikeMe.com (over 155,000 patients and over 1,000 conditions), in which a patient can begin to identify with others who have the same condition, who are undergoing the same treatment or who have the same symptoms. There are also general disease sites like the Association of Cancer Online Resources (ACOR.org), which supports 127 specific cancer communities, and TuDiabetes.org that has communities for those with type 1, type 2, and other forms of diabetes (e.g., late autoimmune, gestational). Some of these sites have been started by concerned family members (e.g., PatientsLikeMe and Children with Diabetes) and have evolved as they sought to sustain and grow their communities. PatientsLikeMe is supported through the sale of data (fully disclosed), while Johnson & Johnson became the owner of Children with Diabetes.[27, 28]

Some healthcare providers promote the use of establishment of social media communities by patients. For example, Children's Hospital Boston (Children's) and many other hospitals provide support for patients to connect with family and friends via CarePages.[29] Social patient support groups at Children's provide a forum in which patients are able to share experiences regarding various treatments at the hospital and learn about specific hospital divisions.[30] Children's even built an app that allows members of the TuDiabetes community to connect, share, and compare their diabetes data around the world.[31]

Other types of social media community use by consumers include "crowdsourcing" their condition—when they may post their own or a family member's symptoms and ask, "Has anyone seen this before?" This may occur in conjunction with setting an appointment with a provider, depending on the seriousness of the condition. From the patient's point of view, this may feel—and be—more efficient than calling all of their friends on the telephone and asking the same question.

Consumers also communicate fitness, diet, and weight loss information with each other via various forms of social media. A growing number of mobile fitness apps have their own social media communities, allowing users to compare walks and runs, or even find someone to go bike riding with. The MapMyFITNESS community now boasts over 8 million members.[32] France-based Withings has created scales, blood pressure cuffs, and baby monitors that can post data to

Twitter, Facebook, a Personal Health Record, or to others using the devices (a community).[33] As described in Dr. Zaroukian's case study in this book (Case Study 23), some physicians are prescribing these social fitness apps, and even enrolling alongside their patients to both monitor and encourage them. Weight Watchers offers communities for support of members to post successes, get support from others, exchange recipes and fitness tips, or ask questions.[34]

Choosing/Rating Providers, Treatments, and Insurers (Patients)

As indicated earlier in this chapter, a large number of consumers (41%) access social media to find, or make decisions about, a provider. A slightly larger number, 42%, report accessing consumer reviews on treatments (12%), doctors (11%), hospitals (10%), or health insurers (9%).[35] Only 17% actually write reviews of physicians.[36] Writing reviews takes more effort than reading and may be driven by the emotion of either a very positive or negative experience. Tara Lagu, MD, of Tufts University School of Medicine performed an analysis that indicates that most of the reviews are positive.[37]

An article in *The Progressive Physician* indicates there were over 30 sites rating providers as of mid-2011.[38] Such sites include Yelp.com, HealthGrades.com, RateMDs.com, Vitals.com, and Dr.Score.com. The *American Journal of Medical Quality* reports it is easier for consumers to find rating sites featuring patient anecdotes and experience than those based on evidence-based measures, which are often run by community organizations such as Minnesota Community Measurement's HealthScores (www.mnhealthscores.org) or government organizations such as CMS' Hospital Compare (www.hospitalcompare.hhs.gov).[39] Some rating sites, such as Vitals.com, encourage physicians to log into their rating page, verify office information, post a photo, and even provides materials for providers to share with patients, encouraging them to post comments.[40]

Patients use a variety of other social media functions, such as "checking in" via Foursquare (mobile app) to share their location with family or friends, and to view tips that others may have left regarding the location (e.g., "Hallways are cold; bring a jacket."). According to healthcare social media monitor Ed Bennett, only 10% of hospitals had "claimed" their Foursquare locations, which enables them to obtain analytics on the visitors, provide promotions, etc.[41] Using a range of social media monitoring tools (both free and paid), some providers are actively monitoring popular social media platforms for any mention of their organization's name—and intervening when a negative patient experience is revealed with the hope of turning the situation around.

What Do Patients Want to Do with Social Media?

The PwC HRI survey queried patients regarding healthcare provider services they would find valuable, if offered via social media, with the following results:

- 72% - availability of doctor appointments
- 71% - appointment reminders

- 70% - referral to specialists
- 69% - discounts or coupons for services
- 69% - continued support post-treatment/discharge
- 68% - voice complaints/seek customer service
- 68% - patient reviews of doctors
- 68% - treatment reminders
- 65% - current ER wait times[42]

HOW ARE PROVIDERS USING SOCIAL MEDIA?

Ed Bennett's resources indicate hospitals are increasingly using social media, and aforementioned surveys indicate physicians are ahead of hospitals in this use.

In the fall of 2011, 1,229 hospitals had a total of 4,118 social media sites, including:

- 575 YouTube Channels
- 1,068 Facebook Pages
- 814 Twitter Accounts
- 566 LinkedIn Accounts
- 946 Foursquare (most unclaimed)
- 149 Blogs

Managing Online Reputation

It is clear from the plethora of provider rating sites and unclaimed foursquare accounts that providers have an online reputation, whether they realize it or not, and whether or not they are managing it. We recommend that providers–both hospitals and physicians–take stock of their online reputation, and then engage with social media not only for their reputation's sake, but ultimately for the many other benefits and functions that social media provides to them and to their patients and potential patients.

Communicate with Patients

Keeping standards of professionalism and the Health Insurance Portability and Accountability Act (HIPAA) in mind, providers can mindfully engage with their patients via both more open platforms like Facebook or custom-built applications that can broadly or strictly limit those who can view communications. Physician associations, like the American Medical Association, the Canadian Medical Association, and the UK's General Medical Council–have all issued policy, guidance or draft guidance regarding their members' use of social media.[43, 44, 45] The physician associations reinforce the issues of privacy and professionalism, and recognize the increasingly well-known benefits as well as challenges of social media use by physicians. They universally recommend physicians consider separating their personal and professional uses of social media.[46, 47, 48] On April 28, 2012, however, the Federation of State Medical Boards (FSMB) adopted policies not only recommending that physicians have separate professional and personal social media accounts, but also discouraging physicians from ever "friending" patients from their personal social media accounts.[49, 50]

Jen Brull, MD, an IT-savvy family physician in rural Kansas, however, has chosen to engage with patients both via her practice's Facebook page, as well as her own personal Facebook page (see Case Study 2). Dr. Brull has indicated that her town is small, thus many of her patients are already her friends outside of Facebook. Seeing them at school, church, and around the community, it seems natural for her to be friends with some patients on Facebook. Though it is not a HIPAA violation for a patient to share his or her own protected health information (PHI), Dr. Brull indicates that if a Facebook conversation on either her professional or personal site appears to risk exposure of PHI, the conversation is directed into a secure, one-to-one channel, such as the telephone or secure messaging via the practice's patient portal.[51] (You can read more about Dr. Brull in Case Study 2.)

There are many examples of both hospital and physician professional Facebook, Twitter and other social media pages that are used for connecting with patients and their local communities. Without exchanging HIPAA-protected information, other forms of communication can take place, such as taking polls, issuing and responding to fitness challenges, provider offices tweeting that they are running behind with appointments, and providers responding to generic questions about medication dosages or other generic health questions that can be responded to by including a link to a credible information source.[52,53] Provider professional societies are actively providing advice and tools to guide their members' use of social media tools, such as the American Nursing Association's "Social Networking Principles Toolkit."[54] Even commercial vendors are supporting their customers' social media efforts, as with Cardinal Health's provision of "The Independent Pharmacists Social Media Toolkit" to its customers—sharing best practices and case studies.[55]

Due to the immediacy with which social media is often handled, especially by organizations that patients know have full-time social media staff (usually evident from blogs and frequency of posts), it can become yet another channel for social media-savvy patients or their families to identify appropriate locations for care in an emergency. Matthew Browning did exactly that, reaching out to Emory Healthcare on Twitter when his wife's grandmother's aorta ruptured, and no other hospital in south Georgia would accept a transfer.[56] The tweet read, *"@emoryhealthcare NEED HELP NOW!! Grandma w/ RUPTURED AORTA needs Card Surgeon/ OR ASAP, STAT! can you accept LifeFlight NOW!!?"*[57]

Emory responded quickly with advice and the number for their transfer service:

Ultimately, Browning's wife's grandmother did end up on a life flight, and the resulting Twitter exchange is priceless:

[58] Not every exchange on social media between patients and providers will end up with such a positive outcome, just as with every in-person encounter or telephone call, but it shows the value of the tools. Ultimately, the majority of the communication necessary for the transfer occurred by telephone, but it was uniquely enabled by the patient's family and the provider communicating via social media.

Disaster Communications

Several hospitals have discovered that their early efforts at communicating with their local communities via social media were rewarded when disasters struck and phone lines were too busy. Scott & White Hospital in Temple, Texas used Facebook, Twitter, and YouTube to communicate with their community in the wake of the Fort Hood shootings in 2009. The posts were originally intended to merely extend their communications and "go where the people are," immediately after the shooting. The hospital started by informing the community via Twitter that they were monitoring the situation, eventually adding that the hospital had become a secured environment, asking that blood donations occur at the main hospital only, providing estimated wait times for blood donations, and posting an interview with the chief of surgery on YouTube (with links from other social media platforms).[59] Innovis Health, located in Fargo, North Dakota, also used Twitter and blogs to communicate with their community when their flood disaster struck in March of 2009.[60]

Monitor Patients

Social media functions have also been incorporated with remote monitoring functions in order not only to monitor chronically ill patients, but to use gamification and rewards to increase self-management compliance. Joseph Cafazzo, MD, of the Toronto-based Centre for Global eHealth Innovation, developed an app and system called "Bant" (featured in Case Study 5), which enabled adolescent type 1 diabetic patients to take their blood glucose readings with a wireless glucometer and smart phone, while communicating with other users via a private and secure, Twitter-like microblog.[61] Points were accumulated for consistent reporting and redeemed for games and music on iTunes. The daily average frequency of blood glucose measurement went up 49.6%, and a large percentage of the children indicated they wanted to continue the program.[62]

Educate Patients

Perhaps one of the most popular reasons for the use of social media is to educate patients and the general public. Examples include:

- **Physician Blogs.** Blogging either for their own practices or on behalf of hospitals, physicians address issues ranging from patient safety to immunization promotion, to a variety of health and wellness topics. Some physician bloggers gain an audience much wider than their own patients, often attracting other physicians and healthcare professionals as followers. Physicians that have gained a broad following include:

  o Wendy Sue Swanson, MD: http://seattlemamadoc.seattlechildrens.org/

  o Bryan Vartabedian, MD: http://33charts.com/

  o Howard Luks, MD: http://www.howardluksmd.com/

  o Mike Sevilla, MD: http://www.familymedicinerocks.com/family-medicine-rocks-blog/

  o Natasha Bugert, MD: http://kckidsdoc.com/

  o Ted Eytan, MD: http://www.tedeytan.com/

- **YouTube Videos.** As indicated earlier, there are 575 hospital YouTube channels. Across industries, videos are fast becoming a preferred educational tool. It is easy to find educational videos such as pre-op education, or general overviews on anything from foot health to eye health to telestroke programs.[63] YouTube educational videos are also being produced and promoted by the Centers for Disease Control and Prevention (CDC), state and local departments of health, and health information organizations.

- **Twitter.** As shared in the section under patients' education, Twitter is often a source that patients turn to for education. Provider and hospitals can share links to articles, events, and other resources in short snippets to those that follow them on Twitter, or by hashtagging such tweets, making them accessible to a broader audience interested in the topic.

- **Facebook Pages.** Facebook remains by far the most popular social networking platform in the world, with an estimated 1 billion members. Physician and hospital Facebook pages can not only provide information about the practice or facility, but posts can be educational regarding conditions, epidemics, self-management, nutrition and fitness—pointing either internally to the provider's own educational resources or providing links to external, evidence-based articles and videos.

These are just a few of the more popular channels used by providers to educate patients.

Marketing: Is There a Return on Investment in Social Media?

Hospital marketing experts Chris Bevolo and Chris Boyer both promote the idea of establishing measures and equations to determine the return on investment (ROI) of providers' social media use.

Chris Bevolo defined it as follows: "Social Media ROI reflects any other marketing-related ROI: the net financial revenue to the organization from the effort, after having accounted for the effort's costs."[64]

Chris Boyer, Director of Digital Marketing and Communications at Inova Health System, speaks often on the topic of ROI, and recommends use of a customer relationship management (CRM) program to enable this tracking. Boyer said, "We use a CRM database to track patients' engagement along the path from social media interactions, newsletter opt-ins or event signups, to becoming a patient (and eventually, generating revenue)."[65]

Pediatrician Natasha Bugert, MD, estimates her practice gains one new patient per week who actively volunteers that they came because they saw her practice's Facebook page or her blog. Dr. Bugert estimates the monetary return as follows:

"Figure 52 patients a year x $2,700 (average pediatric care for 0-24 months) = $140,000 of average billable income over two years."[66]

But Dr. Bugert recognizes there is much more than a monetary return for her participation in social media. Summarizing the outcome of just one year on social media, she stated: "The beauty of social media is that I never talked with these parents about these health and safety issues. Parents made good decisions for their families after getting the information. Period. That's all they needed, and that's all it took."[67]

Provider Education and Other Provider Uses

Many providers believe they are more up to date on developments related to their practices due to their involvement on social media, even as listeners. Jennifer Shine Dyer, MD, reports, "I'm much more well-read now that I'm on social media. Twitter reminds me of skimming journals for the things that interest us."[68]

A variety of sites are set up to facilitate providers connecting with other providers, sharing experiences, and even seeking advice and insights from each other. Sermo, a site exclusively for U.S.-licensed physicians, reports over 125,000 MDs and DOs covering 68 specialties are spending over 35,000 hours per month in physician-to-physician discussions on its platform.[69] Sermo has even launched a mobile app called iConsult, to allow physicians to seek the advice of others in the network regarding diagnoses, with tools for uploading photos and specifying the specialty needed to review the symptoms.[70]

Providers, especially hospitals, are beginning to engage in "enterprise social media" by using private, secure internal communities to communicate asynchronously, but in a manner more collaborative than email, and which may reduce the need for meetings. Some believe that enterprise social media will be expanded to include coordination of care needed especially for patient-centered medical homes or accountable care organizations.[71]

Some of the more "out of the box" uses of social media include the "crowdsourcing" that Cleveland Clinic does via the InnoCentive platform. Having set up a "pavilion" within InnoCentive.com, Cleveland Clinic can launch periodic challenges to the public, enlisting them to solve problems, with rewards going to the best answer. For example, from November 2011 to February 2012, the Cleveland Clinic offered $20,000 to the best solution to "Pinpoint the location of airway collapse in sleeping individuals."[72]

PAYERS' USE OF SOCIAL MEDIA FOR MEMBER ENGAGEMENT

Insurance companies like Humana and Aetna are sponsoring online games and virtual game worlds to promote health and wellness. Some of these combine connecting activity-monitoring devices and competing with family and friends for virtual and real-life rewards.[73, 74]

Mindbloom, which offers a free social personal development platform, is creating an enhanced version of its Life Game for Aetna. Mindbloom allows users to choose areas of life that are important to them—from a list including health, relationships, spirituality, career, finances, creativity, and lifestyle—set goals in those areas, and be rewarded for achieving them. But Mindbloom is not just a goal-tracking platform; it combines two equally important elements: (1) Inspiration, allowing users to create online vision boards with music, images (e.g., pictures of family), and quotes that remind the user of their purpose, passion or "why"; and (2) Action, in which plans are put together to achieve goals in each chosen area for personal improvement. Players can connect with friends (including Facebook friends), co-workers, family, and others using the platform and encourage one another.[75] Mindbloom has also launched a related mobile app called Bloom, which both facilitates inspiration through the music and photos already on the mobile devices, and reminders for a range of activities from working out to bringing home flowers for a spouse.[76]

Similarly, Humana has partnered with game developer Ubisoft to create an online social fitness platform called Your Shape: Evolved 2012, including a new version for 2013. Using game platforms such as Nintendo's Wii and Microsoft's Xbox 360 with Kinect, users set goals, and can choose from a variety of pre-programmed interactive activity modules—ranging from cardioboxing to jumprope, yoga, tai chi, varieties of dance and many more (125 activities in 2013 version).[77] By entering height and weight, approximate loss of calories is calculated. The program also has suggested recipes. In 2012, Ubisoft estimated the Your Shape:

Evolved participants burned over 1 billion calories. Participants can share results or challenge friends on Facebook and see how they compare across others in the U.S. and around the world.[78]

But not all payer activity related to social media is planned. Rather, social media is an engagement platform for both the member and payer, and the engagement is as varied as life itself. When Arijit Guha, a graduate student at Arizona State University, ran up against a $300,000 lifetime limit on his Aetna plan while undergoing chemotherapy for Stage 4 colon cancer, he decided to engage Aetna CEO Mark Bertolini on Twitter.[79] The heated exchange resulted in calls to Guha to discuss his case, and ultimately Bertolini committed Aetna to meeting Guha's medical bills through the end of the year. Tremendous visibility was given to this use of social media, and Guha ended the discussion by committing to contribute the over $100,000 already donated by others (to his cause) toward other cancer patients in need of care.[80]

FUTURE

The future of social media use in healthcare will be a combination of currently known uses diffused in the usual manner (i.e., innovators, early adopters, early majority, late majority, and laggards), as well as new uses evolving (e.g., care coordination), and new platforms emerging.[81] The key is to stay in touch with the consumer uses and trends, and determine the best methods and channels to engage a specific patient population. It is hoped that effective engagement will lead to better outcomes and a healthier population.

Leading an industry-wide effort to effectively use and promote social media through the Mayo Clinic Center for Social Media, the Mayo Clinic stated its own social media philosophy as follows:

"Mayo Clinic believes individuals have the right and responsibility to advocate for their own health, and that it is our responsibility to help them use social media tools to get the best information, connect with providers and with each other, and inspire healthy choices. We intend to lead the healthcare community in applying these revolutionary tools to spread knowledge and encourage collaboration among providers, improving healthcare quality everywhere."[82]

Over 100 healthcare organizations, including many hospitals and IDNs such as Dignity Health, HCA and Bon Secours, as well as the American Hospital Association and Healthcare Information and Management Systems Society (HIMSS), have joined the Social Media Health Network, led by the Mayo Clinic. This network provides tools, resources, and guidance from fellow network members regarding what works, what does not, and what may be emerging.

Providers can monitor social media developments best by becoming actively engaged on social media themselves. Lee Aase of the Mayo Clinic Social Media Center advised: "Take some baby steps and get into the shallow end of the pool as soon as possible. The great strategic ideas will probably come after you have some experience in the social media world."[83]

Lee Aase has also started an informal, free, and fun educational site called the Social Media University, Global (SMUG), which offers step-by-step introductions to social media generally and use of specific tools like Facebook, blogs, podcasting, and even the enterprise social media tool Yammer.[84]

CONCLUSION

Why does social media matter for patients?

Social media benefits patients by meeting them "where they are," allowing them to connect with others to obtain support or education, or to share their own experiences, as well as engaging them via "gamification" techniques designed to make the engagement experience, and adherent behavior, pleasurable.

How does it matter for providers?

The benefits of social media use for providers can include more adherent patients, increased patient acquisition, more efficient patient education (time saved), increased awareness of recent industry developments, and rapid establishment of a secondary communications channel for use during disasters.

Why does it matter for the healthcare system as a whole?

Social media brings highly effective communication channels and functionality into an industry—healthcare—that requires more effective ways for participants to engage with each other with great potential to enhance the provider-patient relationship.

For more information, read the following case studies at the end of this book:

Case Study 2: Innovation in the Context of Meaningful Use

Case Study 3: How Social Media Has Changed My Medical Practice

Case Study 5: Social Media, Mobile and Gamification for Diabetes Care

Case Study 7: Everyone Performs Better When They Are Informed Better

Case Study 10: The E Is for Engagement

Case Study 23: Prescribing Apps for Weight Loss

REFERENCES

1. Bradley A. Defining Social Media: Mass collaboration is its unique value. Weblog entry. Gartner Blog Network. March 8, 2011. http://blogs.gartner.com/anthony_bradley/2011/03/08/defining-social-media-mass-collaboration-is-its-unique-value. Accessed July 7, 2012.

2. Hinchcliffe D. Social business holds steady gap behind consumer social media. ZDNet Enterprise Web 2.0. ZDNet. Web. August 27, 2011. Accessed July 7, 2012.

3. Modahl M, Tompsett L, Moorhead T. Doctors, patients, and social media. *QuantiaMD*. September 2011. Web. http://www.quantiamd.com/q-qcp/doctorspatientsocialmedia.pdf. Accessed July 7, 2012.

4. Bennett E. Three years later. Social Media Resources for Health Care Professionals from Ed Bennett. October 27, 2011. http://ebennett.org. Accessed July 7, 2012.

5. Modahl M, Tompsett L, Moorhead T. Doctors, patients, and social media. *QuantiaMD*. September 2011. Web. http://www.quantiamd.com/q-qcp/doctorspatientsocialmedia.pdf. Accessed July 7, 2012.

6. Wasserman T. Facebook to hit 1 billion user mark in August. *Mashable*. January 12, 2012. http://mashable.com/2012/01/12/facebook-1-billion-users/. Accessed September 29, 2012.

7. Piombino K. Infograph: The cost of marketing on Facebook. Ragan.com. Lawrence Ragan Communications, Inc., December 6, 2011. http://www.ragan.com/Main/Articles/Infographic_ The_cost_of_marketing_on_Facebook__44059.aspx. Accessed July 7, 2012.

8. Shaer M. Game of life. *Popular Science*. February, 2012.

9. Cafazzo JA. Innovation in the Medical Home: How mobile and social technologies can accelerate health behavior changes. Patient Centered Primary Care Collaborative. Webinar. January 26, 2012.

10. Boyer C. The ROI of social media. In: *The Thought Leaders Project: Hospital Marketing* (Kindle Edition). Minneapolis: Bierbaum Publishing, LLC. 2011; pp 491-492.

11. Healthcare Information and Management Systems Society (HIMSS). Healthcare 'friending' social media: What is it, how is it used, and what should I do? December 2, 2012. www.himss. org/asp/contentredirector.asp?contentid=79496. Accessed July 7, 2012.

12. Shine Dyer J. Physician Panel. Health Care Social Media Summit. Mayo Clinic. Rochester, MN; October 17, 2011.

13. Holtz S. Hospital Social Media Best Practices. Health Care Social Media Summit. Mayo Clinic. Rochester, MN; October 7, 2011.

14. Holtz S. FIR Interview: Steve Widmann and Aaron Hughling, Smith and White Healthcare. NevilleHobson.com. January 23, 2010. Podcast. January 31, 2012.

15. Social media 'likes' healthcare. PricewaterhouseCoopers, LLP. April 2012. Web. http://www. pwc.com/us/en/health-industries/publications/health-care-social-media.jhtml. Accessed July 7, 2007.

16. Fox S. The Social Life of Health Information, 2011. Pew Research Center's Internet and American Life Project. May 5, 2011. Web. July 7, 2012.

17. Social media 'likes' healthcare." PricewaterhouseCoopers, LLP. April 2012. Web. http://www. pwc.com/us/en/health-industries/publications/health-care-social-media.jhtml. Accessed July 7, 2007.

18. http://www.symplur.com/healthcare-hashtags/. Accessed December 23, 2012.

19. Torrey T. Bio. About.com. Patient Empowerment. Web. http://patients.about.com/bio/Trisha-Torrey-35320.htm. Accessed December 23, 2012.

20. Torrey T. Welcome to the patient empowerment forum. About.com Patient Empowerment. Web. http://forums.about.com/n/pfx/forum.aspx?nav=messages&webtag=ab-patients&lgnF=y. Accessed December 23, 2012.

21. Sparling K. Six Until Me. http://sixuntilme.com/. Accessed December 23, 2012.

22. Society for Participatory Medicine. Web. August 25, 2012. www.participatorymedicine.org.

23. *Journal of Participatory Medicine*. Web. August 25, 2012. http://www.jopm.org/about/. Accessed December 23, 2012.

24. Halamka J. The Life of a Healthcare CIO. December 2011. http://geekdoctor.blogspot. com/2011/12/we-have-cancer.html. Accessed December 23, 2012.

25. Social media 'likes' healthcare. PricewaterhouseCoopers, LLP. April 2012. Web. http://www. pwc.com/us/en/health-industries/publications/health-care-social-media.jhtml. Accessed July 7, 2007.

26. http://www.facebook.com/pages/Shingles/105819746124450. Accessed December 23, 2012.

27. How we make money. PatientsLikeMe.com. http://www.patientslikeme.com/help/faq/ Corporate#m_money. Accessed July 8, 2012.

28. About children with diabetes. Children With Diabetes, Inc. Web. http://www. childrenwithdiabetes.com/about.htm. Accessed July 8, 2012.

29. For patients and families. Children's Hospital Boston. Web. http://childrenshospital.org/patientsfamilies/Site1393/mainpageS1393P150.html. Accessed July 8, 2012.

30. Social media 'likes' healthcare: Social studies – Patients are using social media to better educate themselves; page 9. PricewaterhouseCoopers, LLP. April 2012. Web. http://www.pwc.com/us/en/health-industries/publications/health-care-social-media.jhtml. Accessed July 7, 2007.

31. TuDiabetes and Children's Hospital Boston launch online effort to map U.S. diabetes metrics. Children's Hospital Boston. May 29, 2010. Web. http://childrenshospital.org/newsroom/Site1339/mainpageS1339P630.html. Accessed July 8, 2012.

32. About. MapMyFITNESS. Web. http://www.mapmyfitnessinc.com/company/. Accessed July 8, 2012.

33. www.withings.com. Accessed December 23, 2012.

34. Weight Watchers. Web. http://www.weightwatchers.com/. Accessed August 25, 2012.

35. Social media 'likes' healthcare: How Consumers Are Using Social Media; page 5. PricewaterhouseCoopers, LLP. April 2012. Web. http://www.pwc.com/us/en/health-industries/publications/health-care-social-media.jhtml. Accessed July 7, 2007.

36. Ibid.

37. Atoji C. Understanding online physician rating sites. *The Progressive Physician*. May 10, 2011. Web. http://theprogressivephysician.com/practice-management/understanding-online-physician-rating-sites.html. Accessed July 8, 2012.

38. Ibid.

39. O'Reilly KB. Patient Rating websites top Google Searches for best doctors. *American Medical News*. The American Medical Association. January 28, 2011. Web. http://www.ama-assn.org/amednews/2011/11/28/prsb1128.htm. Accessed July 8, 2012.

40. Atoji C. Understanding online physician rating sites. *The Progressive Physician*. May 10, 2011. Web. http://theprogressivephysician.com/practice-management/understanding-online-physician-rating-sites.html. Accessed July 8, 2012.

41. Bennett E. Hospitals on Foursquare – A National Survey. Social Media Resources for Health Care Professionals from Ed Bennett. February 6, 2011. Web. http://ebennett.org/hospitals-on-foursquare/.

42. Social Media 'Likes' Healthcare: How Consumers Are Using Social Media; figure 8, page 16. PricewaterhouseCoopers, LLP. April 2012. Web. http://www.pwc.com/us/en/health-industries/publications/health-care-social-media.jhtml. Accessed July 7, 2007.

43. AMA Policy: Professionalism in the Use of Social Media. American Medical Association. Web. http://www.ama-assn.org/ama/pub/meeting/professionalism-social-media.shtml. Accessed July 8, 2012.

44. Social Media and Canadian Physicians: issues and rules of engagement. Canadian Medical Association. Web. http://www.cma.ca/advocacy/social-media-canadian-physicians. Accessed July 8, 2012.

45. Doctors' Use of Social Media: A Draft for Consultation. General Medical Council. Web. http://www.gmc-uk.org/Draft_explanatory_guidance___Doctors_use_of_social_media.pdf_48499903.pdf. Accessed July 8, 2012.

46. AMA Policy: Professionalism in the Use of Social Media. American Medical Association. Web. http://www.ama-assn.org/ama/pub/meeting/professionalism-social-media.shtml. Accessed July 8, 2012.

47. Social Media and Canadian Physicians: issues and rules of engagement. Canadian Medical Association. Web. http://www.cma.ca/advocacy/social-media-canadian-physicians. Accessed July 8, 2012.

48. Doctors' Use of Social Media: : A Draft for Consultation. General Medical Council. Web. http://www.gmc-uk.org/Draft_explanatory_guidance___Doctors_use_of_social_media. pdf_48499903.pdf. Accessed July 8, 2012.

49. Model policy guidelines for the appropriate use of social media and social networking in medical practice. Federation of State Medical Boards. Web. http://www.aafp.org/online/en/ home/publications/news/news-now/opinion/20120629editbrull.html. Accessed July 8, 2012.

50. Press Release: New FSMB policy addresses appropriate use of social media by physicians. Federation of State Medical Boards. May 2, 2012. Web. http://www.fsmb.org/pdf/nr-social. pdf. Accessed July 8, 2012.

51. Healthcare Information and Management Systems Society (HIMSS). Healthcare 'friending' social media: What is it, how is it used, and what should I do? December 2, 2012; page 16. www.himss.org/asp/contentredirector.asp?contentid=79496. Accessed July 7, 2012.

52. How doctors are using social media to connect with patients. *US News & World Report.* November 21, 2011. Web. http://health.usnews.com/health-news/most-connected-hospitals/ articles/2011/11/21/how-doctors-are-using-social-media-to-connect-with-patients. Accessed July 8, 2012.

53. Brull J. Social Media in Medicine: Do your patients 'like' you? American Academy of Family Physicians. June 29, 2012. Web. http://www.aafp.org/online/en/home/publications/news/ news-now/opinion/20120629editbrull.html. Accessed July 8, 2012.

54. http://www.nursingworld.org/socialnetworkingtoolkit.aspx. Accessed July 8, 2012.

55. http://pharmacy.about.com/b/2011/04/04/113.htm. Accessed July 8, 2012.

56. Emory Healthcare. Can Twitter Help Save Lives? A Health Care Social Media Case Study, Part 1. Posted April 27, 2011. Web. http://advancingyourhealth.org/highlights/2011/04/27/can-twitter-help-save-lives-a-health-care-social-media-case-study-part-i/. Accessed September 16, 2012.

57. Ibid.

58. Ibid.

59. Holtz S. FIR Interview: Steve Widmann and Aaron Hughling, Smith and White Healthcare. NevilleHobson.com. January 23, 2010. Podcast. Accessed January 31, 2012.

60. Video case study: Using social media, A communication tool during a natural disaster: Russell Herder. Web. http://vimeo.com/4668036. Accessed July 8, 2012.

61. Cafazzo JA. Innovation in the Medical Home: How mobile and social technologies can accelerate health behavior changes. Patient Centered Primary Care Collaborative. Webinar. January 26, 2012.

62. Ibid.

63. Swedish Telehealth Program. Swedishseattle YouTube Channel. December 15, 2011. Web. http://www.youtube.com/watch?v=P65n0h0f6w4. Accessed July 8, 2012.

64. Bevolo C. *A Marketer's Guide to Measuring Results: Prove the Impact of New Media and Traditional Healthcare Marketing Efforts.* Marblehead: HCPro; 2010.

65. Boyer C. The ROI of Social Media. In: *The Thought Leaders Project: Hospital Marketing.* (Kindle Edition). Minneapolis: Bierbaum Publishing, LLC; 2011; pages 491-492.

66. Bugert N. How social media has changed my medical practice. KevinMD.com August 2011. Web. http://www.kevinmd.com/blog/2011/08/social-media-changed-medical-practice.html. Accessed December 23, 2012.

67. Ibid.

68. Shine Dyer J. Physician Panel. Health Care Social Media Summit. Mayo Clinic. Rochester, MN. October 17, 2011.

69. Discuss + Connect. Sermo.com. Web. http://www.sermo.com/take-a-tour/discuss-connect. August 25, 2012.

70. Sermo and Real-Time Medicine™ Bring You 'Diagnosis on Demand. YouTube.com. http://you tube/ksDOc4Mklx4. Accessed August 25, 2012.

71. Ferguson G. Personal Interview; July 6, 2012.

72. InnoCentive Challenges. InnoCentive.com. Web. https://www.innocentive.com/ar/challenge/browse. Accessed July 9, 2012.

73. HG4H. Humana Games for Health. Web. http://www.humanagames.com. Accessed January 22, 2012.

74. Health IT News Staff. Aetna encourages health through online gaming. *Healthcare IT News*. March 4, 2011. Web. Accessed January 22, 2012.

75. Mindbloom. What's the Life Game? Web. https://www.mindbloom.com/?learn_more. Accessed September 15, 2012.

76. Mindbloom. What is Bloom? Web. http://www.mindbloom.com/about/bloom./ Accessed September 15, 2012.

77. Best Buy. Your Shape: Fitness Evolved 2013 – Nintendo Wii U. Web. www.bestbuy.com. Accessed September 15, 2012.

78. Your Shape: Fitness Evolved 2012. Game Features. Web. http://yourshapegame.ubi.com/fitness-evolved-2012/en-US/game-features/index.aspx. Accessed September 15, 2012.

79. Hensley S. Cancer patient gets help from 'Bake Sale" and Aetna CEO. *National Public Radio*. July 30, 2012. Web. http://www.npr.org/blogs/health/2012/07/30/157591196/cancer-patient-gets-help-from-bake-sale-and-aetna-ceo. Accessed September 16, 2012.

80. Huffington Post. Guha A. Cancer patient wins Twitter war against Aetna Insurance. July 31, 2012. Web. http://www.huffingtonpost.com/2012/07/31/arijit-guha-cancer-patient_n_1724478.html#slide=1322205. Accessed September 16, 2012.

81. Rogers E. Diffusion of innovations. Stanford University. Web. http://www.stanford.edu/class/symbsys205/Diffusion%20of%20Innovations.htm. Accessed July 9, 2012.

82. About. Mayo Clinic Center for Social Media. Web. http://socialmedia.mayoclinic.org/about-3/. Accessed July 8, 2012.

83. 6 tips from Lee Aase about getting involved with social media. Ragan's Health Care Communication News. April 4, 2010. Web. http://www.health carecommunication.com/Main/Articles/6_tips_from_Lee_Aase_about_getting_involved_with_s_5049.aspx . Accessed July 9, 2012.

84. Social Media University, Global. Web. http://social-media-university-global.org/. Accessed August 25, 2012.

Chapter 10

Privacy and Security Challenges of Improved Patient Engagement

By Adam H. Greene, JD, MPH, and
*Deven McGraw, JD, MPH, LLM**

INTRODUCTION

Evolving to a more patient-centered healthcare system that actively engages patients depends on more robust information access and communication between providers and patients (and their caregivers). Innovative technologies like electronic health records (EHRs), mobile applications, e-mail, texting and social networking provide opportunities to enhance information access and provider-patient communication, with the potential for enormous benefits to individual and population health. Consumers are using these innovative tools for nearly all other aspects of their personal and professional lives, so it is not surprising that there is an increasing desire on the part of patients to use these tools to promote and sustain health and well-being. However, providers are understandably concerned about their ability to use these tools while protecting the privacy and security of sensitive health information.

Most healthcare providers are obligated to follow the Health Insurance Portability and Accountability Act (HIPAA) Privacy and Security Rules with respect to sharing identifiable ("protected") electronic health information with patients. Patients have a right to access and receive electronic copies of digital health information in EHRs; at the same time, providers have obligations under the

---

*The authors thank Helen Ovsepyan, Davis Wright Tremaine LLP, for her invaluable contributions to this chapter.

Privacy and Security Rules to safeguard that information and prevent inappropriate or unauthorized access to it. Providers may also have data protection obligations under the Confidentiality of Alcohol and Drug Abuse Patient Records Regulations (42 CFR Part 2) and state health data privacy and security laws. This chapter focuses on how providers can responsibly share information digitally with patients using technology while maintaining compliance with HIPAA.

As noted throughout this chapter, **HIPAA does not prohibit providers from using technology to share data and communicate with patients.** In fact, with respect to some provider-patient communications, the opposite is true—patients have a right under the Privacy Rule to receive copies of their health information upon request. However, HIPAA does obligate providers to take certain measures to ensure such communications take place in a way that (1) meets the needs of the patient; and (2) ideally, minimizes risks to patient privacy. This chapter discusses how providers can meet their HIPAA obligations and still effectively use technology to engage patients.

Patients are not obligated to comply with HIPAA or other health privacy laws; however, the sensitivity of health information does not disappear when the information is collected and shared by patients. Privacy and security are a shared responsibility, and patients should be aware of the potential risks of sharing their health information in certain contexts. Patients who are educated about such risks— and about more privacy-enhancing ways to store and share their health data—will be more empowered to actively use technology to support their health goals.

GENERAL COMPLIANCE ISSUES

Common Elements for Patient Engagement in Compliance with HIPAA

HIPAA should not stop healthcare providers from effectively communicating with their patients. Rather, there are certain compliance measures that providers should keep in mind when engaging in any type of patient communication—whether by email, text messaging, etc.—that will help providers steer clear of the potential privacy and security pitfalls in these areas.

Compliance with the Security Rule—Risk Assessment, Risk Management and Implementation of Appropriate Safeguards

In order to ensure that communications with patients are in line with the Security Rule, a healthcare provider must institute or, at a minimum, address a number of standards and implementation specifications with respect to administrative, physical, and technical safeguards of electronic protected health information.[1] It is important to remember that HIPAA does not just govern diagnostic information; the definition of "protected health information" governs almost any individually identifiable health information that relates to the provision or payment of healthcare.[2]

Some of the Security Rule's implementation specifications are required (such as conducting a risk analysis), whereas others are "addressable" (such as encryption of transmitted protected health information). For addressable implementation specifications, the entity must document whether the specification is reasonable and appropriate and, if so, must implement the specification.[3] If, with respect to any given electronic protected health information, an addressable implementation specification is not reasonable and appropriate, then the provider should document why not and implement an equivalent alternative measure if reasonable and appropriate.[4]

The following are a number of key requirements under the Security Rule that providers must keep in mind when contemplating electronic patient engagement[5]:

- Conduct an accurate and thorough risk analysis regarding the potential risks and vulnerabilities to the confidentiality, integrity, and availability of electronic protected health information created by the patient engagement.[6] The provider may consider that different types of messages (e.g., patient-specific treatment information versus preventive tips distributed to the entire patient population) may carry different levels of risk based on different levels of harm if disclosed to unauthorized persons;

- Based on this risk analysis, implement security measures sufficient to reduce risks and vulnerabilities to a reasonable and appropriate level[7];

- Implement appropriate technical security measures to guard against unauthorized access to electronic protected health information that is being transmitted over an electronic communications network and (addressable) implement a mechanism to encrypt electronic protected health information where appropriate[8]; and

- Develop and maintain reasonable and appropriate policies and procedures, training, and safeguards with respect to electronic communications to patients, such as to guard against sending communications to the wrong recipient (e.g., potential human error such as typos when entering email or text message addresses).

When implementing any such safeguards, providers should evaluate and balance competing concerns. **After all, the primary reason for patient engagement is to promote patient information-sharing. While the security of patient information is important, the Security Rule offers enough flexibility to permit providers to address security concerns while still effectively communicating with patients.** The key will be whether the provider has conducted a risk analysis, appropriately managed that risk, and documented the process.

Compliance with the Privacy Rule—Honoring Patient Preferences

When implementing appropriate safeguards, providers should also be cognizant of patient preferences. Specifically, the Privacy Rule gives patients the right to receive communications of protected health information from the provider by alternate means or at alternate locations when reasonable.[9] This means that a patient can make reasonable requests to be contacted at different places or in a different way. For example, a patient may request that appointment reminders are sent via unencrypted email, rather than through postcards. **The provider must accommodate**

reasonable requests for such alternative forms of communications.[10] Providers should be cautious about allowing security measures to trump reasonable patient preferences, such as by informing a patient that the provider will not communicate according to the patient's preferences due to concerns of interception by third parties. Rather, providers may be well advised to ensure that the patient is aware of any risks and then accommodate any reasonable requests. An exception may be where the patient's request creates unreasonable risk to other patients' information, such as because the requested form of communication could introduce malware into the provider's information systems.

TECHNOLOGY-SPECIFIC HIPAA SECURITY ISSUES

As further discussed above, healthcare providers must consider a number of compliance issues when electronically communicating with patients, regardless of the type of technology. In addition, there are technology-specific issues that healthcare providers should take into account. The following sections address these technology-specific security issues.

Mobile Devices

Sometimes key moments of electronic patient engagement do not happen with the clinician sitting behind a desktop, but rather occur through that clinician's use of mobile devices. Providers may wish to include such mobile devices in their risk assessment and risk management plan, including personal devices that may be used to maintain or transmit messages to patients. When using mobile devices such as smart phones to communicate with patients, providers should consider risks such as loss, theft, unauthorized access, malware, and cloning.

After taking steps for risk assessment and management, providers can consider appropriate administrative safeguards in accordance with 45 C.F.R. § 164.308. This may include policies on when it is or is not appropriate to communicate with patients through personal and provider-owned mobile devices, appropriate training on such policies, and appropriate sanctions for violating such policies. Where reasonable, 45 C.F.R. § 164.310 calls for physical safeguards needed to keep mobile devices physically secure (e.g., procedures for disposal and reuse of electronic media and maintaining an inventory of hardware and media containing patient information). Finally, 45 C.F.R. § 164.312 provides for a number of technical safeguards. For example, providers may wish to consider appropriate authentication of mobile devices to ensure that only appropriate persons can view any patient information stored on the devices,[11] and appropriate encryption of the devices should they become lost or stolen.[12] Many smart phones include built-in encryption capabilities for data at rest.[13]

Email

Email is frequently used to communicate with patients. **Rest assured; email is a permissible form of patient engagement when used in accordance with the Security Rule.**[14] As discussed above, the first steps in using email for patient communications is to assess and manage the risk. After that, the provider must employ reasonable and appropriate administrative, physical, and technical safeguards. Providers may consider adjusting controls based on the risk associated with each type of messaging and document the decision-making process and conclusions. For example, the risks and appropriate safeguards corresponding to appointment reminders for a general care hospital may differ from those corresponding to treatment-specific messages of an oncologist.

Communications by email will pose a risk to electronic protected health information that is at rest (e.g., emails in a provider's inbox or outbox and resting on servers), as well as electronic protected health information in transit (e.g., emails as they are being sent or received). One way to manage risk for emails, whether at rest or in transit, would be to implement a technical safeguard such as encryption, or to avoid placing sensitive information into the body of the email but instead send the patient a notification that a message is waiting on a secure website (i.e., a portal through which secure messaging may occur). The appropriateness of such safeguards may vary based on the nature of the health information.

Administrative safeguards relevant to emailing include having a business associate agreement with any entity that uses or discloses electronic protected health information on the healthcare provider's behalf (e.g., an Internet-based email provider).[15] For email that is being transmitted, a common risk is human error (e.g., where staff enters the wrong email address), especially with public email servers where any transposed characters will likely lead to an email going to the wrong party (rather than the email being returned as undeliverable). Possible safeguards to address this may be policies or procedures requiring staff to double-check emails or use a patient address book where the emails have been confirmed by sending verification emails. Physical safeguards may include physical security for email servers and any devices that retain copies of the email (e.g., copies of email on work stations, smartphones, or other portable devices). Providers may also wish to explore and perform a risk assessment relative to use of the relatively new Direct secure messaging protocol, or another secure messaging system provided by their EHR vendor.[16]

Texting

While the government has issued some guidance on emailing patient information, the risks and guidance surrounding text messages is less clear. This makes it more difficult for healthcare providers to assess and manage risk under the Security Rule. Because there are no rules or guidance on means to safeguard text messaging, it is even more important to conduct and document a well-reasoned risk analysis and assessment.

Text messages may not have end-to-end encryption during transmission, and encryption that is employed by telecommunications carriers for wireless transmissions may be weak and subject to unauthorized decryption.[17] As a result, healthcare providers should not ignore the possibility that text messages containing electronic protected health information might be intercepted by unauthorized persons, but should recognize that such attacks may have a low probability due to the level of expertise necessary for such interception.

As with email or any other use or disclosure of protected health information, what are reasonable and appropriate safeguards will depend on the nature of the information you intend to transmit via text message and the risk associated with each type of messaging. In addition to the safeguard methods noted above for emailing (such as, in the case of texting, ensuring that the recipient's phone number is correct), safeguards for text messaging may include using technology such as secure messaging software instead of SMS texting.

As is true for the use of mobile devices and e-mail to communicate with patients, providers should also factor in the needs and desires of their patients with respect to text messaging. Patients can be advised of the potential risks, and providers should do what they can to mitigate those risks—but for many patients, the convenience of receiving appointment reminders or other notifications by text message may outweigh the remaining risks. Providers may want to develop policies with respect to text messaging (for example, what types of information are eligible to be sent by text message), share these policies with their patients and document agreement by patients to receive text messages in accordance with these policies.

Social Media

Social media is a form of communication that is becoming increasingly popular. However, there are some major hurdles to provider use of social media in compliance with HIPAA. Because the social media providers may retain copies of the messages being posted, social media may be a non-starter unless a healthcare provider can obtain a business associate contract with the social media provider storing the messages on the healthcare provider's behalf. For example, if a patient sends a message to a healthcare provider through Facebook or Twitter, the healthcare provider may not be able to respond through those same platforms if the response includes protected health information (e.g., relates to the patient's care) in the absence of a business associate agreement with the social media provider, since sending such a response may mean having the social media provider maintain the message on the healthcare provider's behalf. Rather, the healthcare provider should consider returning the message through other means. Additionally, by replying to "friend" requests from patients, providers may be disclosing protected health information unless they have the proper privacy settings in place. Accordingly, healthcare providers should be sensitive to whether accepting friend requests may expose a patient relationship in violation of HIPAA or other confidentiality obligations. Finally, even if a provider organization does not formally use social media, the provider's staff may. Accordingly, it is important to caution employees against posting any information about patients on their social media pages. Even if they believe they are posting anonymous information, a posting may be considered to identify a patient under HIPAA simply because the date of service is readily ascertainable.

Portals

Increasingly, healthcare providers are offering patients the ability to become more engaged in their care through Internet-based portals, either to improve patient satisfaction and interaction or as part of the Heath Information Technology for Economic and Clinical Health (HITECH) Meaningful Use incentive program. Whatever the reason, the availability of a patient portal introduces privacy and security issues that need to be addressed.

As with the use of other technologies, providers deploying patient portals will need to conduct a risk assessment and formulate a risk management plan that implements appropriate and reasonable safeguards. This may include encryption of the information in transit and at rest, authentication of the patient, and logging and review of system activity. Healthcare providers should consider that certain security measures, such as robust password requirements, may act as an obstacle to patient engagement and should consider whether risks can be addressed through a combination of less obtrusive controls.

One significant issue relates to minors, specifically with respect to providing patient portal access to third parties such as parents or caregivers. Providers have different options for providing such third-party access under HIPAA. For example, healthcare providers can provide access to parents as personal representatives without the need to obtain an authorization.[18] However, such access may need to be immediately cut off at the age of majority, and the provider may not provide access to information about certain healthcare services for which the minor can consent under state law (such as certain reproductive services). For minors that are of an age where they can consent to certain services under state law, a provider may choose to only provide portal access to the minor's records if both the minor and parent or guardian provides an authorization.

Blue Button

The Blue Button functionality was developed to provide patients with the ability to log into a secure website to view and have the option to download their health information or claims information. The Blue Button program has been implemented by several federal government agencies such as the U.S. Department of Veterans Affairs, Medicare, and TRICARE. As of August 2012, one million veterans had registered to access their health information via the VA's Blue Button functionality.[19] Several other public and private sector organizations are also looking to offer similar download capabilities for their patients and members, and the industry is looking to create ways for patients to forward downloadable record sets to other physicians. The Blue Button program is evolving as a means for patients to not only download data, but also to share the data with their healthcare providers. For example, a software application is available that allows a patient to push the patient's Blue Button data directly to the mobile device (e.g., iPad) of the patient's clinician.[20]

If a healthcare provider accepts Blue Button data from a patient, the provider should be sensitive to the corresponding privacy and security issues. These include the security considerations for mobile devices that have been identified above. Additionally, if the provider relies upon the Blue Button data for treatment, then the provider should be sensitive to the corresponding privacy obligations, such as that the information may be subject to the individual's rights of access and amendment and the information may need to be included in documentation of the location of the provider's designated record sets.[21] Providers also may want assurance of the source of information shared by a patient—did the information come from a provider EHR, without alteration, or is the information generated by the patient?

Video Conferencing and Telemedicine

With developing technology, video conferencing and telemedicine are increasingly useful in providing patient care, especially in underserved geographic areas. However, as with social networking, where information about the patient contact or the actual video interaction will be stored by the third-party provider, a business associate agreement may be necessary. If the service provider is not willing to enter into such an agreement, HIPAA compliance may become a major hurdle.

Healthcare providers should consider whether the video conference or telemedicine session will be encrypted in transit (many software applications include this feature). Additionally, if the application stores any patient information on the device, the provider should address any corresponding risks.

CONCLUSION

Healthcare providers have legal and ethical obligations to safeguard electronic, identifiable health information and handle it with care and sensitivity. The requirement to protect patient privacy is often put forward as an obstacle to the more active use of innovative technologies to improve communications between patients and providers. But in most cases, HIPAA need not be an obstacle to greater patient engagement using technology. However, myths and uncertainty about the application of current health privacy and security laws to the use of technology by patients and providers has the potential to slow their adoption to improve individual and population health.

It is critical for patients and providers to understand that in most cases, current law does not present an obstacle to the use of technology for communication of health information between providers and patients. Providers must perform a risk assessment and implement reasonable risk mitigation and management strategies prior to implementation of these technologies. This chapter includes a number of tips for providers seeking to successfully balance mitigating risk and communicating more effectively with their patients. Patients also need to understand that HIPAA does not provide protections for health information that they collect and share and that they are responsible for educating themselves about the data sharing policies of any mobile or Internet-based tools that they decide to use for health purposes.

Why are privacy and security protections important to patients?

Safeguarding the privacy and security of patients' data is important to build the trust of patients in sharing their confidential data in order to gain important benefits for health and well-being. Without that trust, many patients will be reluctant to use innovative technology for self-management and to engage with their providers. Even patients who are comfortable sharing (or even publically posting) personal health data have a right to expect that their providers will protect it on their behalf and leave the choice about whether or not to share it in less secure environments in their hands.

Why are privacy and security protections important to providers?

Providers have obligations under HIPAA to pay attention to risk mitigation and management strategies they use when communicating with patients via electronic media and using personal health information technology. It is equally important for providers to understand that these obligations should not be a barrier to communicating electronically.

Why are privacy and security protections important to the healthcare system as a whole?

As it becomes the norm for patient data to be electronically exchanged between providers and between providers and patients, privacy and security rules ensure that the data can be trusted to be accurate and secure. This increases the reliability of all data in the healthcare system, with corresponding potential to increase patient safety.

For more information, read the following case studies at the end of this book:

Case Study 6: Kaiser Permanente—My Health Manager

Case Study 21: Patient Portals Offer Opportunities for Patient-Provider Interaction at Children's Hospital of Philadelphia

Case Study 22: Veterans Health Administration, Department of Veterans Affairs—Patient Engagement and the Blue Button

REFERENCES

1. 45 C.F.R. §§ 164.308, 164.310, and 164.312.
2. 45 C.F.R. § 160.103 (definition of "protected health information").
3. 45 C.F.R. § 164.306(d)(3).
4. *Id.*
5. The Security Rule provides dozens of standards and implementation specifications. However, not all requirements are created equal (first and foremost comes risk analysis and management), and not all will be relevant or appropriate to the topic of patient engagement. This chapter does not undertake the task of addressing each and every one of these standards and implementation specifications. Rather, we focus on those that are most relevant to the issue of electronic patient engagement and the various technologies involved.
6. 45 C.F.R. § 164.308(a)(1)(ii)(A).
7. 45 C.F.R. § 164.308(a)(1)(ii)(B).
8. 45 C.F.R. § 164.312(e)(1) and (2)(ii).
9. 45 C.F.R. § 164.522(b).

10. HHS Office for Civil Rights. Does the HIPAA Privacy Rule permit health care providers to use e-mail to discuss health issues and treatment with their patients? http://www.hhs.gov/ocr/privacy/hipaa/faq/health_information_technology/570.html. Accessed September 3, 2012.

11. 45 C.F.R. § 164.312(d).

12. 45 C.F.R. § 164.312(a)(2)(iv).

13. Belfort RD, Ingargiola SR, Deven McGraw D, et al. Strategies for safeguarding patient-generated health information created or shared through mobile devices. *Lessons from Project HealthDesign;* February 9, 2012. Additionally, many smart phones have the capability for users to add third-party applications such as encryption, decryption, and other security tools to their devices.

14. Does the HIPAA Privacy Rule permit health care providers to use e-mail to discuss health issues and treatment with their patients, supra note __.

15. A business associate contract is not required for parties acting merely as a conduit as opposed to using or disclosing PHI as part of their services. *See* HHS Office for Civil Rights: Are the following entities considered 'business associates' under the HIPAA Privacy Rule: U.S. Postal Service, United Parcel Service, delivery truck line employees and/or their management? March 14, 2006. http://www.hhs.gov/ocr/privacy/hipaa/faq/business_associates/245.html.

16. The Direct Project. "The Direct Project Overview." October 11, 2010. Available at http://wiki.directproject.org/file/view/DirectProjectOverview.pdf. Accessed December 23, 2012.

17. The interception of text messages is rare, but not unprecedented. Nusca A. Code that encrypts world's GSM mobile phone calls is cracked. ZD Net, December 28, 2009. http://www.zdnet.com/blog/btl/code-that-encrypts-worlds-gsm-mobile-phone-calls-is-cracked/28942?tag=mncol;txt. Accessed December 23, 2012.

18. 45 C.F.R. § 164.502(g).

19. U.S. Department of Veterans Affairs. Blue Button reaches one million registered patients. August 31, 2012. http://www.va.gov/opa/pressrel/pressrelease.cfm?id=2378. Accessed December 23, 2012.

20. Humetrix. Using the iBlueButton® Professional iPad App in compliance with HIPAA. http://www.humetrix.com/ibbpro_hipaa_compliance.html. Accessed December 23, 2012.

21. 45 C.F.R. §§ 164.524, 164.526.

Chapter 11

The Role of Design in Patient Engagement

By Aaron E. Sklar and David Fetherstonhaugh

INTRODUCTION

The success of the healthcare industry is increasingly dependent on engaging people in their health. What does engagement mean? An engaged person's daily life is consistent with a healthy lifestyle. An engaged person recognizes when a health issue arises and gets into action to address it. An engaged person is tapped into a network of people and resources in his or her community who can provide help. An engaged person is empowered to make informed decisions regarding health options.

Healthcare providers are adopting more and more tools, practices, technologies and services that contribute to empowering and engaging people in navigating their health actions and outcomes. Many of the other chapters in this book serve to demonstrate the value of the various tools and approaches emerging to engage people in their health. As a field, we are building a body of evidence around an ever-increasing toolkit of capabilities that enhance patient engagement. Now we are at the tipping point, ready to take the next step into the engagement realm—using design as our vehicle.

We are not the first industry to go down this path. We can look at how other industries evolved from functional offerings to more polished—more human-centered—offerings. The winners in consumer product industries have moved beyond comparing functionality and are competing—and winning—based on the design of their offerings. They are tapping into human needs and desires beyond "engagement" and going so far past functionality as to generate product lust.

Similarly, the winners in the healthcare arena will be those who take that next step to designing products, services, and experiences that speak to authentic human needs and desires. In the very near future, patient engagement will be the price of entry in a competitive marketplace and those who succeed will be the ones who go beyond the basics and connect with people using tools and services that fit into our everyday lives with ease and delight.

BEDSIDER CASE STUDY: DESIGNING FOR ENGAGEMENT

For decades, our society has sought to teach people how to prevent unplanned pregnancy. There is an abundance of information around established best practices for abstinence and birth control methods, yet getting people engaged and in action on this topic has been a challenge, as shown by our continued high rates of unplanned pregnancies. According to a recent Health and Human Services (HHS) study, 37% of U.S. babies are a "surprise."[1] The National Campaign to Prevent Teen and Unplanned Pregnancy has made major headway in enlisting people to take action, and their secret weapon has been design as a tool to get the message through. Their Bedsider service (www.bedsider.org) provides birth control information and motivation in a truly engaging way. The tone speaks to women about sex in terms that resonate and uses mobile technology and social media in a unique way that fits into the lives of the participants. For example:

"Compared to men from 30 years ago, today's typical American male has a smaller sperm count. But they can still get you preggers, so take your pill."

"Nearly 1/3 of all Italians have had sex in a garden. That certainly puts the dirt in dirty. Take your pill now. Stop to smell the roses (and have sex) later."

　– Bedsider.org birth control text reminders

The Bedsider pilot studies have demonstrated remarkable participation rates and affirming positive feedback from the women who are benefiting from the services—and much of their success is due to the design strategy. The design team began with in-depth ethnographic research in-context with women who are struggling with birth control practices. Given the topic, these in-context conversations got personal, very quickly. The insights derived from these interviews guided the selection of the right tone of voice (not too smutty, not too medical), as well as the best technology approach. Through a series of rapid prototyping efforts, the design team landed on the use of text messages as reminders for women who were taking a daily birth control pill. The tone of the messages is playful and provocative. The reminders are different each day and are about something people care about (better sex) instead of something they do not (birth control). The team knew they were onto something when they discovered that many of the women in the pilot study were forwarding the text messages onto friends and family.

The Bedsider initiative highlights the role of design in engagement. By taking traditionally sterile educational information and transforming it into a lively and fun conversation between peers and communities, the impact is powerful—and the outcome is extraordinary. In the first pilot study, at intake, almost all women (99%)

reported that Bedsider sounded like a useful site. Of these women, 70% went on to use Bedsider. After a three-month pilot, over 90% of participants reported that Bedsider helps lessen fears about side effects of birth control and/or helps them to use contraception more accurately. Over three-quarters (80%) of the participants who used Bedsider at three months believed that since they started using it, they have been trying harder to avoid unprotected sex. Additionally, the percentage of doctor appointments kept over time increased for Bedsider users while decreasing for those women who did not use Bedsider.[2] These results are unprecedented in the arena of engaging women in birth control.

Other examples of positive outcomes through the use of patient-centered design principles are detailed in several of this book's case studies, specifically itemized at the end of this chapter. These studies highlight the way that design principles can be applied both to designing technology and to designing (and refining) processes and operations that put patients at the center.

APPLYING ESTABLISHED DESIGN STRATEGY APPROACHES TO HEALTH

Design is a fundamentally different approach to engagement. It is optimistic and exploratory. While design is intrinsically creative and intuitive, there is an established process and structure that works and can be strategically applied to challenges in the health domain. We all can recognize the role design plays in the physical world. The same approach and design processes are critical in creating interactions and experiences that engage people in services, experiences and even processes. Learning from engagement strategies used in successful consumer product industries, we can apply the same principles to engage people in health. Three core characteristics underlie all design strategy: discovery, empathy and prototyping.

Discovery—Diverge and Converge

Design starts with an open mind, asking "What if..." and "How might we..." The optimistic tone generates dozens if not hundreds of ideas. There are many practices aimed at divergent thinking to explore the opportunity space fully—such as brainstorming, analogous inspiration, and system mapping. These expansive methods are paired with convergent practices to keep the process moving forward—synthesis, visualization, and reviews to name a few. Alternating between divergent and convergent practices continually expands the playing field of options and then focuses the effort.

Empathy—Lead with People

People provide the inspiration, feedback, and purpose necessary to drive good design decisions. Empathy is achieved through immersion into the lives and context of potential users. Touch points with people permeate any successful design effort. Initial connections and continued encounters with potential end-users are the foundation for all of the ideation that follows. In addition to commonly used techniques like focus groups and surveys, design teams explore dozens of ways of engaging with people in an empathetic manner, including ethnographic studies, shadowing techniques, and behavioral probes. Structured feedback sessions guide a constant learning process to help evolve ideas into solutions that will resonate and result in the intended outcomes.

Prototyping—Always Be Iterating

Design places an enormous emphasis on prototypes as a vehicle for learning quickly and cheaply. Rapid prototyping techniques yield dozens of prototypes at every stage of an initiative. Prototypes can be physical, virtual, or experiential. They can be made of materials on-hand such as paper, glue, video, or even a skit. Regardless of the format, a continuous cycle of production, feedback, and refinement is critical to sustained success. Rather than building toward a proof of concept at the end of an initiative, the design approach is to create dozens of small prototypes throughout—by the time a team is ready for a pilot study, elements of the concepts have already been validated by potential users and iterated upon several times.

Design challenges come in all shapes and sizes—regardless of the context, the same core process is effective. Within the healthcare domain, one obvious setting for applying these practices is personal health IT tools. The same emphasis on discovery, empathy and prototyping is equally important when developing a patient-centered workflow or launching an outreach campaign. As provider organizations work to design patient-friendly and patient-centered processes, design strategies can help ensure that the results achieve desired outcomes. Perhaps the biggest opportunity for impact is in applying design processes to transformative personal health IT tools that fundamentally change how and why people engage with technology in the context of health.

Design is a powerful tool in our campaign to impact public health by enabling people to make healthier choices, take their medicines more regularly, and engage in healthy behaviors.

DESIGNING FOR BEHAVIOR

With engagement in health as a focus, the role of designing for behaviors takes a prominent role. Discussed next are behavior design principles that will surface in any patient engagement activity. Underlying each principle is a fundamental human need—satisfying these needs is what keeps people engaged. Not meeting these needs, however, makes it more likely that people will disengage out of confusion, frustration, or disappointment.

Discovery: Put the Offer on the Path the Patient Is Already On

The first step of the engagement journey is about helping people discover that a service is available that can meet their need. At this point in the journey, people may only have a rudimentary understanding of what their problem is. This stage of the journey requires the designer to understand what potential users are already doing and place invitations where they will be discovered in the natural flow of experience.

Opportunity: Shape Your Offer to Fit the Patient's Circumstances

Once the user has discovered that a service might be relevant to his or her life, the next stage of the journey is to help find the right fit for the user's particular circumstances. The more flexible your offer is, the easier it is to remove any barriers to participation. A simple example of this is an appointment scheduling system that accommodates patients who cannot make personal phone calls during working hours. In this example, offering evening or online scheduling creates opportunities for engagement. The Stanford Persuasive Technology Lab has demonstrated the effectiveness of another way to fit into people's circumstances: connecting new behaviors you want to encourage to existing behaviors that people are doing already.[3]

Navigation: Build Momentum Through Small Steps Along the Journey

Once you help users find the fit that works for their circumstances, the next stage of the journey is to help them along the path that initiates the service itself. Similar to a car GPS system that provides turn-by-turn directions, any tools that help patients navigate the journey will increase engagement. Reminder phone calls or text messages within 24 hours of an appointment are a simple example of this principle.

Reinforcement: Provide Feedback to Integrate New Behaviors

Once users begin engaging in the service itself, they begin the process of integrating new behaviors into their existing routines. This is a ramp-up period in which users are learning new systems and encounter failure that may require assistance. Any opportunity for feedback—both positive and negative—will enhance engagement and build new patient behavior muscles for the future. The best kind of feedback allows for learning to occur. Provide feedback in a way that prompts the new behaviors you want to encourage.

Evolution: Sustain Behaviors over Time

Once new routines are established, users may need help staying on the path. Over time, a person's motivations may change or he or she may become habituated to the cues that helped trigger behaviors initially. The design team must walk a fine line between creating habits that can become automatic routines and offering variety to keep engagement high. The Bedsider example does a good job of keeping the same format of a spicy text message every day at the same time, while mixing up the content of the message to keep people interested enough to read it each day.

DESIGN'S SECRET SAUCE

Engagement is not a new topic. Many industries have successfully built relationships with people, and engaging people in their own health has amazing potential to have a huge impact on public health, as well as the sustainability of the healthcare industry. Health, while impactful, requires a fresh look at engagement. People do not care about their health–they care about living their lives fully; thus health is a means to an end. It is up to the healthcare industry to do the work to discover what it is that people care about and how they can connect in relevant ways. Design provides both a process to learn and a creative mindset to reframe health-related product and service offerings in ways that resonate. Leveraging design and design thinkers in patient engagement will open up a promising new realm of conversations between empowered patients and providers.

CONCLUSION

Why does design matter for patients?

The core principle of empathy places patients in the center of every design process. Patients are the source of inspiration and continuous feedback to shape the final outcome in a way that ensures engagement.

Why does design matter for physicians?

As patient engagement becomes the price of entry in an increasingly competitive landscape, design can become the powerful tool that distinguishes physicians' offerings from one another.

Why does design matter for the healthcare system as a whole?

As a society, we have enjoyed amazing breakthroughs in medical technology. As we continue to make strides in technology and scale, design is a vehicle for achieving the health outcomes we are all committed to through its unique role at the intersection of people and technology.

For more information, read the following case studies at the end of this book:

Case Study 5: Social Media, Mobile and Gamification for Diabetes Care
Case Study 8: From God to Guide
Case Study 14: Moving the Point of Care into Patients' Daily Lives via Text Reminders
Case Study 15: New Models of Education for Pediatric Diabetes
Case Study 16: Online Therapy and Patient Engagement
Case Study 20: Good Days Ahead, Online Cognitive Behavioral Therapy

REFERENCES

1. Mosher WD, Jones J, Abma JC. Intended and unintended births in the United States: 1982-2010. National Health Statistics Report. Number 55; July 24, 2012.

2. The National Campaign to Prevent Teen and Unplanned Pregnancy. The Case for Bedsider. March 2011. www.thenationalcampaign.org/Bedsider/Case-for-Bedsider.pdf. Accessed September 30, 2011.

3. Fogg BJ. Tiny Habits, Persuasive Technology Lab, Stanford University. Posted December 30, 2010. http://www.slideshare.net/captology/3-steps-to-new-habits. Accessed September 30, 2010.

Chapter 12

The Future of Patient Engagement

By Brad Tritle, CIPP

INTRODUCTION

To precisely predict the future of patient engagement would require the ability to read everyone's mind, as well as predict future linkages between great ideas, technical advances, market needs, capital, and policy. As this is not possible, we will do the next best thing: look to forward-thinking ideas that appear to be working in pilots or in limited scope that—based on what we know of market needs, capital and policy—have the capability to scale.

We will spend the first part of this chapter looking at the market needs and trends, capital (primarily reimbursement and business models), and policy. The second part of this chapter will look at some ideas that are working in limited fashion with the potential to scale—ideas in motion. Finally, we will briefly address what may be needed for some of these ideas to take hold, including provider education, encouragement, leadership, and policy changes.

MARKET NEEDS/TRENDS, CAPITAL AND POLICY

Increased Consumerism

Overall, patients have more healthcare and wellness-related products and services to choose from than ever before. Whether it is the ability to choose between allopathic, homeopathic, naturopathic and other forms of medicine to find relief for a condition, to choose one of the 13,000 healthcare related apps referred to in Chapter 8, to chat online with those who have a similar diagnosis to learn about treatment options, or to go online to research and choose a provider based on consumer reviews or government ratings, patients have many options—and they discover many of them through the Internet.

The Internet has changed many other industries, in some cases facilitating the elimination (disintermediation) or downsizing of professions that stand between consumers and information or services. Today, when you make travel arrangements, do you still call a travel agency or do you go online directly either to an airline or value-added (consolidated choices or discounts) website? Or did you download an airline's or other site's mobile app? In some cases, information itself is the desired end product. In other cases, the information allows the consumer to make choices about an end-product or service (e.g., researching a car online before going to a dealership to test drive and buy).

The ease with which consumers can access information about other aspects of their lives has raised consumers' "information expectations" in healthcare. Susannah Fox at the Pew Research Center has shown that 59% of adults go online in search of health information.[1] Over one million veterans using the VA's Blue Button proves that patients want access to their own health information.[2] The implication of these expectations is that information that *should be* available–*is* available (transparency).

Having more information–*faster*–also seems to foster ownership and enhanced decision making outside of healthcare. As referenced in Chapter 11, this builds engagement behavior "muscles" that patients begin to use in other aspects of healthcare.

What is "consumerism" as applied to healthcare? McKinsey and Company, in a report published in the *American Journal of Managed Care,* described healthcare consumerism as "an orientation to new care delivery models that encourage and enable greater patient responsibility through the intelligent use of information technology."[3]

Consumers' interest in this use of technology is confirmed by many of the findings in the 2011 Deloitte Center for Health Solutions' survey of 4,000 Americans:

- 66% of consumers stated they would consider switching to a provider who offers access to medical records through a secure Internet connection.[4]

- 61% expressed interest in using a medical device that enables them to both check their condition and send information to their provider through a computer or cell phone.[5]

- 52% stated they would use a smart phone or similar device to monitor their health if they could access their medical records and download information about their condition and treatments.[6]

Consumerism or empowerment is a "mega trend" driving more "demand-control" in healthcare–a recognition that by effectively facilitating patients' management of their health and wellness, the incidence of costly chronic disease and condition treatment may decrease.[7] The Centers for Disease Control (CDC), for example, has proven that for every $1 spent educating a diabetic patient about self-management, $8.76 in healthcare cost savings occurs.[8]

But what other trends or factors will impact the future of patient engagement?

Value-based Care/Reimbursement

The very act of changing reimbursement methods from fee-for-service to a model that pays for outcomes, keeping people as healthy as possible, and the care coordination to make that happen will require greater communication with patients. It is estimated that 80% of heart disease, stroke, and type 2 diabetes could be prevented if the modifiable chronic disease risk factors of physical inactivity, smoking, and unhealthy diet were changed. The patient maintains the greatest control over those factors.[9] Thus, it will financially benefit providers to have their patients engage in effective self-management of their health and wellness. Increased movement toward value-based reimbursement methodologies—such as the patient-centered medical home (PCMH) and the accountable care organization (ACO) should result in a great desire and need for patient engagement. In fact, as noted in Chapter 2, several of the models have built-in requirements to create loyalty and engagement. For example, the joint principles for PCMH are to include patients and families in quality improvement activities at the practice level, and the American Academy of Family Physicians (AAFP) provides recommendations for patient satisfaction surveys and patient focus groups.[10] Medicare-based ACOs, in fact, are required as part of the Accountable Care Act to "promote evidence-based medicine and patient engagement, such as through the use of telehealth, remote patient monitoring, and other such enabling technologies."[11]

Consumer-Directed Health Plans (CDHP)/High-Deductible Plans

Due to the continuing increase in health insurance costs for employers, more continue a shift to consumer-directed health plans (CDHP), which often are high-deductible health plans and which may be coupled with health savings accounts (HSAs). It is believed that asking the consumer to take a greater financial role in their healthcare will motivate them to take more responsibility for their health and wellness. As indicated by a Rand Corporation study, employers may choose to support employees using CDHPs by providing decision support tools—supporting both financial and healthcare decisions—through "enhanced personal health records," or "personal health platforms."[12] As consumers come to their doctors with questions about both the cost and efficacy of different treatment options, we can imagine new shared decision models that take into account values about the price and cost of care, as well as level of intervention and treatment efficacy.

Value-based Insurance Design (VBID)

Value-based insurance design is related to CDHP in that value-based insurance plans also ask consumers to take more financial responsibility for their health. They approach this in a more nuanced way than CDHP plans, however. Consumers in a value-based plan may pay a greater share of their healthcare costs, but only for what the plan considered to be "lifestyle" or "optional" care (e.g., knee replacement or services thought to be of less "clinical" value). At the same time, consumers are motivated by paying a smaller share or nothing for things that will have a more dramatic impact on their overall health such as getting annual physicals or taking cholesterol or diabetes medications. Under these plans, co-pays might be lowered, and diabetes medications may be offered free to motivate patient adherence to treatment plans. The incentives are in themselves a form of engagement, but it is obvious that spending more for compliance could also include investing more in engagement technology.

Practice Business Models

A variety of new business models for practices are being explored. PCMH and ACO-related participation by providers was described above, and earlier in this book, in Chapter 6, direct primary care (DPC) was briefly mentioned. Again, providers experimenting with, or committed to, DPC are attempting to "do more with less," and may end up leapfrogging other practices in their use of patient engagement technologies in order to promote efficiencies resulting from enhanced communication.

Other practice models include the private physician or concierge medicine model. This is similar to the DPC, except that typically there are fewer patients per physician, and those patients pay a higher annual fee, usually with no additional charges for office visits. Most concierge models include use of a standard electronic health record (EHR), and many include a variety of integrated physician and patient tools, including patient education and secure messaging.

Both the DPC and concierge models emphasize greater physician-patient communication, and physician availability to the patient. Growth in these models would promote growth in supportive engagement tools, and related innovations.

Meaningful Use Stages

As was indicated in Chapter 3 on Meaningful Use, Stage 2 has expanded the use of patient engagement capabilities (by both providers and patients), such as secure messaging. Stage 3 is expected to continue this expansion. The "Theme" for Stage 3 is "Patient Access to Self-management Tools." At the August 1, 2012 meeting of the Health IT Policy Committee, the following Stage 3 recommendations were discussed (see Chapter 3 for additional information):

- THEME: Patient access to self-management tools.
- Provide 50% of patients the ability to designate to whom and when (i.e., auto blue button & on-demand) a summary of care document is sent to specific care team members (across settings/providers), and create ability of providers to review/accept updates.
- Provide 10% of patients with ability to submit information (types of information, semi-structured questionnaire, home monitoring devices).
- Allow 10% of patients to update/correct information.
- Add (non-English) language support to educational materials.
- Measure: More than 15% of patients use secure electronic messaging to communicate with eligible professionals (EPs).
- Record patient preferences for communication (EPs, not hospitals; >20%).

Though it is too soon to know the outcome, some have discussed the possibility of continuing Meaningful Use with a Stage 4 and Stage 5. If so, these stages would undoubtedly continue to expand engagement of patients and families.

Lean Healthcare/Need for Administrative and Financial Efficiencies

A trend in many healthcare organizations is to implement "lean" principles. Lean is simply defined as "maximizing customer value and minimizing waste."[13] Though lean principles are derived from the Toyota Production System for manufacturing automobiles, they are applicable across all industries. A key perspective of lean is to look at an entire system for opportunities to improve efficiencies, generally through removing bottlenecks. Such improved efficiencies often result in greater flow, which often means more efficient use of resources, cost savings, and a better experience for the customer, as we discussed in Chapter 7.

A straightforward example of these principles in healthcare is available in the prescribing and prescription renewal process for medications. Surescripts estimates that when paper scripts are used, physician offices and pharmacies spend a great amount of time on the phone either calling to confirm the paper script, or calling to obtain or provide a renewal authorization.[14] In fact, MGMA's Group Practice Research Network estimated that the unnecessary administrative complications related to the paper prescription process cost the average practice $15,700 annually per full-time physician and that a practice of ten physicians encounters $19,444 worth of waste each year confirming drug formulary issues by phone with pharmacies.[15]

E-prescribing, on the other hand, eliminates the ambiguities of written prescriptions, and provides decision support tools for the physician regarding formularies, dosages, drug interactions, and more. The prescription is sent directly to the pharmacy, often eliminating any wait time for the patient (who often can pick it up directly), and resulting in more first-time prescriptions being filled. If renewals are authorized, patients can request a new prescription from the pharmacy online and the pharmacy can interact with the physician if a new authorization is required. Use of an electronic platform for communication that includes the provider, payer (drug formularies), pharmacist, and patient eliminates system inefficiencies, saving time and costs for all involved.

Such lean principles can be applied to other existing areas of inefficiency in provider offices, such as requiring patients to arrive early to fill out paperwork that they might otherwise fill out at home online the evening before, and then having provider office staff rekey this same information into the practice management system or EHR when the patient's online entry could suffice and just be checked for completeness. Opportunities to apply lean principles to healthcare abound, and result in both more efficient organizations and happier participants on both sides. Continued application of lean will result in further use of engagement tools, and more capacity in the healthcare system to handle the forthcoming tsunami of demand, as the number of Americans aged 65 years or older is expected to double over the next 38 years.[16]

Personalized Health

Advancements in genomics are occurring daily, increasing the speed of genome mapping, decreasing the costs, and identifying biomarkers that indicate risks. As a result, individuals will be empowered with more information about their bodies than ever before. Though many believe that physicians may be in the driver's seat regarding application of genomics to medicine, Eric Topol, MD, author of *The Creative Destruction of Medicine*, said that he agrees with physician/entrepreneur Hugh Reinhoff, MD, that "Doctors are not going to drive genetics into clinical practice. It's going to be consumers…"[17, 18]

Announcements made in the fall of 2012 dramatically reduced the time necessary to get usable genomic data back to someone for use in treatment. These advances, as well as the likelihood that the cost will drop over time, make it possible to imagine a moment when most medications and most treatments will be customized to patients' genomic profile to ensure the most effective results.

The capabilities of ambient, wearable sensors that can send streams of biometric data to analytic engines raise the possibility that in the future, patients and their physicians will be able to understand and predict events based on a level of granular data that has never before been available. Imagine an asthmatic whose breathing patterns and peak flow levels could be combined with other data such as external temperature, pollen, hormone, and medication levels to understand triggering factors in a whole different, personalized way. Many patients might want someone else to analyze their data for them, but the most motivated may find new ways to analyze their data and contribute to an understanding of their diseases—and their health.

Some patients who currently have implanted devices such as implantable cardioverter/defibrillators (ICDs) are lobbying for access to the data from the devices, arguing that it is their bodies that are generating the data and they should have access to it. We think this will become a non-issue in the future, as patients and providers alike—not to mention medical device manufacturers—will accept patients' right to their data without argument.

Continuing/Unforeseen Technology Developments

We have shared a variety of technology developments in this book that have come to market recently, and the speed of ongoing technological development is astounding. It seems as though predicting developments should be analogous to simply executing a math equation. For example, "When we can all finally agree on A, then A will be added to B and we'll have C in the marketplace." Or, "When enough consumers want it [A], then someone will create the tools [B], and we'll have that type of engagement [C]." In reality, though, developments often take us by surprise.

We must not forget that the future of patient engagement will be influenced by technology developments outside of healthcare, many that we have not anticipated. This argues the need for everyone in healthcare to be alert to developments outside of the field that can be applied to healthcare.

Regulatory and Licensing Changes

There are always opportunities to either encourage or stifle innovation and engagement via regulatory or licensing changes. These can occur at a municipal, state, or national level.

Examples include:

- Privacy laws (e.g., The European Union has a Data Protection Directive; the U.S. has the Health Insurance Portability and Accountability Act [HIPAA]; some U.S. states require more consumer consent than required by HIPAA). Requirements for stronger consumer consent have necessitated informing patients, which is a way of engaging them. In 2002, the original consent requirement in the 2001 HIPAA Privacy Rule was removed in order to lessen the burden on providers. We speculate that this might actually have had the unintended consequence of reducing patients' knowledge of and interest in what is happening with their health information. This same issue—required consent as an opportunity to engage—should be considered as states consider Health Information Exchange legislation.

- Telemedicine laws (14 states have telehealth parity laws—mandating insurance companies to reimburse for televisits that are equivalent in function to in-person visits). By passing these laws, providers can begin engaging patients more easily at home (or while mobile) in ways that are known to be reimbursable. These laws should have a dramatic impact on consumer engagement, though many believe Medicare taking more dramatic steps on telehealth reimbursement would automatically cause other health plans to follow—and Medicare is not impacted by these state laws.

Presidential and Congressional Elections/Healthcare Reform

We should stress the point here that health IT is a bipartisan or non-partisan issue. In fact, in relation to promoting consumer-mediated exchange, think tanks representing both parties supported a bill in Congress in 2007 called the Independent Health Record Trust Act, which would have established patients as the drivers of their own electronic health information exchange (HIE) through new bank-like institutions.

Similarly, as referenced in Chapter 3 regarding Meaningful Use, political leaders such as Representative Patrick Kennedy (Dem), former Speaker Newt Gingrich (Rep), former Senate Majority Leader Bill Frist, MD (Rep), Hillary Clinton (Dem), Paul Ryan (Rep), HHS Secretary Kathleen Sebelius (Dem), former HHS Secretary Michael Leavitt (Rep), former President Bush (Rep) and President Obama (Dem) have all expressed high interest in the use of health IT, with former President Bush calling for all Americans to have an electronic personal health record by 2014, and President Obama signing the HITECH bill as part of the American Recovery and Reinvestment Act (ARRA).[19]

One of the last acts of former HHS Secretary Leavitt (under former President Bush) was to establish a Medicare personal health record (PHR) Choice Pilot program, with four commercial PHR companies participating, and also to establish the Fair Information Practice Principles (which encourage engagement) as part of a toolkit for guiding health IT and PHR development.[20] Similarly, HHS Secretary Sebelius (under President Obama) has further established the need to clearly communicate with patients, in fact pushing for patients to have better access to their own health information, as she stated: "When it comes to healthcare, information is power. When patients have their lab results, they are more likely to ask the right questions, make better decisions and receive better care."[21]

We say this to make the point that no matter how future congressional or presidential elections go, there will be support for both health IT and consumer engagement. The primary difference between parties may be the willingness or extent to which tax dollars are used for related initiatives.

Most of the current incentives (or potential penalties) relative to patient engagement are not tied to the Healthcare Reform/Affordable Care Act (ACA). Thus, even if Healthcare Reform were repealed at some future point, incentives for Meaningful Use (established in the ARRA HITECH stimulus bill) will move forward, and with them the associated incentives for patient engagement.

As discussed earlier in this chapter, several ARRA-created initiatives, such as ACOs, PCMHs, and DPC will encourage more patient-centered care and more patient engagement.

An ACA-specific initiative is Medicare's Hospital Readmission Reduction Program, which provides specific penalties to hospitals that have high readmission rates within certain Diagnosis Related Groups (DRGs). Many hospitals have launched or are launching remote patient monitoring and other programs to enable engagement of the patient via mobile or home-based devices (e.g., scales, glucometers, pulse oximeters), and performing necessary interventions (when indicated by device data) via nurse-led call centers or home health. A readmission typically has meant more revenue for a provider, though poorer quality of life for the patient; the ACA-based readmission reduction program and its financial penalties along with other population health initiatives are tipping the scale in favor of remote (away-from-hospital) engagement that results both in lower costs and higher quality of life for the patient.

Economy

The current poor economy can be seen as impacting consumer engagement in many ways. A few examples include:

- Employers concerned with healthcare costs and lower productivity (due to illness and/or doctor office visits) are incorporating tele-visit and e-visit services in their employee benefit programs. Most are for emergent or urgent care services and are designed to provide follow-up information back to the employee's primary care physician.

- According to the Deloitte consumer health survey, the economic downturn has caused consumers to stay away from healthcare providers when they otherwise would have sought care.[22] Perhaps engagement of such impacted consumers through personal health IT tools could help identify when a visit is truly necessary, provide alternatives to in-person visits that may even be reimbursable by health insurance in some states, and help maintain patient loyalty between actual visits.

- One in four (25%) consumers state they decided to not see a provider when sick or injured; 1 in 5 (20%) delayed or skipped treatment recommended by a provider. The rate of both of these behaviors has increased, with cost-related reasons cited by 53%.[23]

- Three out of four (75%) consumers state the downturn has impacted the amount they are willing to spend on healthcare services and products.[24]

IDEAS IN MOTION

Many exciting ideas are moving patient engagement forward. The concepts discussed next are either proven, or the predictable outcomes of today's trends.

Accessible and Interconnected Data

One need only look at consumers' access to their financial data to see where healthcare is going. One significant current trend is to provide patients with easier access to their health information stored in disparate locations, such as portals from providers, state immunization registries, pharmacies, laboratories, and health information exchanges (HIEs).

In fact, the federal government's initiatives related to Blue Button have not only proven to meet a need (over 1 million veterans downloaded their health information from the Veterans Administration [VA] via the initial iteration of the Blue Button), but are spawning a nationwide movement.[25] Blue Button was initially a simple method for veterans at first and then Medicare recipients to download their health information in a human-readable text format. The initiative has been transferred to the Office of the National Coordinator (ONC), and has become the Automate Blue Button Initiative (ABBI).[26] With a tremendous amount of stakeholder input, it is being expanded to facilitate patients' ability not only to view and download human-readable text, but also transmit the file to a PHR or other patient-endorsed recipient (e.g., primary care provider, family member, home health) in machine-readable, as well as human-readable format. This will enable the data to be automatically "consumed" and presented to another health information system. As referenced in Chapter 3 on Meaningful Use, Blue Button will likely be an option in qualified EHR systems to allow providers to meet some patient engagement Meaningful Use requirements for "View, Download, Transmit" of information by the patient.

One intriguing project related to Blue Button is the attempt to place a Blue Button linked to a statewide HIE inside an employer-sponsored PHR/wellness platform. The Nebraska Health Information Initiative (NeHII), which is a statewide HIE, is working with the SimplyWell's PHR platform to enable employees to have access to clinical information sourced from HIE-participating providers.[27] The HL7 organization's development of a continuity of care document (CCD) to Blue Button transformation tool is assisting in this process, and this tool is expected to be used throughout the country to further enable payer and provider systems to implement Blue Button.

Similarly, the Direct project, described in greater detail below and in Chapter 3, will allow providers not only to communicate with each other but to send information (possibly facilitated by Blue Button) to patients, their PHRs, or other designated recipients via a secure email/messaging protocol.

In the financial world, arrangements have been made between such disparate systems (e.g., bank, investment and credit card accounts) and financial aggregator portals to give consumers a complete picture of their finances, easily accessed and consolidated. Similarly, to give patients, caregivers, families, and providers a complete picture of the patient, it is likely that health information aggregators—such as a community health record, health record bank, or other form of integrated PHR—will become the source of truth for the patient's longitudinal health record. Consumers' ability to share a longitudinal record that they control with providers and others is what is termed "Consumer-Mediated Exchange" by the federal government, and it is one of the three major categories of acceptable HIE.[28] Given the complexity of healthcare information and the opportunity for duplicate information to overwhelm clarity in a PHR, we believe it will not become the norm until tools simplify this process for patients.

New Partnerships with Providers

Through movements such as the Society for Participatory Medicine (S4PM), engaged patients (e-patients), patient advocates, physicians, nurses, family members, and vendors are moving forward a paradigm of medicine as a partnership between patients and providers. On S4PM's website, it describes participatory medicine as "...a cooperative model of healthcare that encourages and expects active involvement by all connected parties (patients, caregivers, healthcare professionals, etc.) as integral to the full continuum of care. The 'participatory' concept may also be applied to fitness, nutrition, mental health, end-of-life care, and all issues broadly related to an individual's health."[29]

The primary cry of the S4PM is to "Give me my dam data," with the word "dam" intentionally implying that patient data has been "dammed up," like a lake created by a dam. By sharing data with patients, not only can lives be saved (as evidenced by many stories of S4PM members, as well as examples in the patient safety chapter), but informed collaboration is facilitated.

Though some of the society's revolutionary ideas may seem radical, they are intended to "stretch the rubberband" of the patients' role and challenge traditional thinking. Proposed ideas include things like patients releasing a Request for Proposal to have providers "bid" on their cancer treatment (decision criteria could include quality and engagement mechanisms, as well as price) or patients sending bills to providers when they have to wait longer than 15 minutes for an appointment may seem radical. These ideas are the symptoms of consumers' overall frustration with their healthcare experiences. Ultimately, we can see that there is a trajectory of increasing expectations by consumers that, once they experience a higher level of engagement, make it difficult for them to return to a provider who doesn't provide enhanced engagement.

This enhanced engagement and partnership, though, is not just so that patients will feel better emotionally. The Chronic Care Model, developed by Ed Wagner, MD, and disseminated across the world, states that improved outcomes proceed from productive interactions between "informed, activated patients" and a "prepared, proactive practice team."[30]

Lucien Engelen described this best in Case Study 8 as the provider changing from a role of "god" to "guide." It may be a difficult transition for some, as years of medical training still cannot ensure that a provider is completely knowledgeable on a moment's notice about the latest evidence-based treatments, and a patient presenting may have just spent hours looking through the latest scholarly journals or reviewing blogs of other patients being treated with the same symptoms. Thus, patients' expectations for their provider to be "all knowing" have been lowered. This does not mean that there is less need for the physician, nor even less respect for physicians, but instead an approach from the patient perhaps saying, "With your medical training and experience, my knowledge of my symptoms, and our jointly looking at my health information and evidence-based options—what should we do? What do you think?"

As Dave deBronkart (known as "e-Patient Dave"), co-founder of the Society for Participatory Medicine and frequent public speaker (also with a case study in this book), stated: "The key in any case is to realize that the world has changed, especially the 'membranes' that used to separate information from the public. Today things evolve so fast, it's not rare—nor a failure—for a patient to have seen something their doctors haven't, as described in Chapter 4. As with any other professional skills development, it's an opportunity for growth in competence."[31]

Quantified Self Movement

Around the world, there is a movement by consumers to be able to measure themselves in various domains—financial, fitness, academic and nutritional. This is also known as the "self-tracking," "body hacking," and "quantified self (QS)" movement.[32] It is not terribly new to track calories or keep a running log, but the new trend gets more granular—how many milligrams of caffeine are consumed, sleeping patterns (via brain wave monitoring of light, deep and rapid-eye-movement sleep), alcohol intake, etc.[33] Though some people doing this are early adopters or fitness buffs, the movement also includes many people battling health conditions and seeking to understand the underlying cause or to prevent recurrence of unwanted symptoms.

Thirty years ago, if you were to ask someone whether they took vitamins, they would say, yes, but today, you may even hear them respond with precise vitamin dosages (e.g., 1,000 mg of vitamin C, 600 mg of calcium). Even without genomic data (which are also desired), patients in the quantified self movement understand that their bodies are unique and may respond differently than their spouses to a certain number of hours of sleep. Those with allergies, in particular, are monitoring pollen counts and specific allergens. Even Ford Motor Company is taking this seriously, enabling a car's Ford SYNC to control smartphone apps, and monitor and display allergens and pollen counts for the car's occupants. This is especially important for those suffering from asthma, as 70% of those with asthma also suffer from allergies of some kind ("Honey, the ragweed is really bad here, I think it would be better for little Johnny's asthma if we stopped for lunch in the next town.").[34, 35] For others, it may not be a health condition, but just a desire to function at their highest capacity during the day. One thing is certain: purveyors of advancements in personalized health, mentioned earlier in this chapter, will find those in the quantified self movement to be ready and willing customers.

When patients have access to more data on themselves and sensors become more passive and less intrusive, they may bring such data with them to an office visit, and providers should be prepared to engage. Some of the data may be surprisingly helpful: can you imagine a patient with cardiovascular disease coming in with an averaged daily blood pressure over the past 30 days, rather than relying on a single reading by a person in a white coat?

Direct Project

In a rather short period of time, it is possible that providers will keep track of patients' Direct secure email/messaging addresses as part of the routine demographics confirmed at each visit (or changeable by the patient through a patient portal). These Direct addresses may correspond to the patient's PHR, remote monitoring hub, or a specific monitoring device, and enable trusted bi-directional exchange of information.

The Direct Project–a U.S. (could be international) standard for trusted secure messaging–enables two known/trusted (authenticated) individuals to communicate. It evolved from a series of blogs by health IT leaders and was identified in the beginning as, at a minimum, a simple and easy way to implement provider-to-provider (one-to-one) HIE in a "push" manner–one provider pushing necessary patient information to another. ONC endorsed the project initially as the Nationwide Health Information Network Direct (NwHIN Direct), and then facilitated community discussions to enable the idea and associated framework to advance. Fairly quickly the idea evolved so that patients could also have Direct addresses, and Microsoft HealthVault was quick to jump on this opportunity for users of their personal health platform. A HealthVault Direct address, for example, could be firstname.lastname@direct.healthvault.com.

By establishing a trusted, known address for the patient, medical records can be securely and easily provided to the patient, even copying family members, home health, or the primary care provider. In fact, one PHR provider, NoMoreClipboard, initiated the concept of a "cc:me"–for patients to be copied when their health information is sent using Direct from one provider to another, or as a way for their data to be sent directly to their PHR.

"One of the barriers to widespread consumer adoption of portable personal health records has been the difficulty of populating a PHR with existing clinical data," said NoMoreClipboard.com President Jeff Donnell. "cc:Me simplifies the exchange of health information in an industry-standard format that is interoperable with any certified electronic health record system used by physicians and hospitals, with no need for integration."

David Kibbe, MD, advisor to the American Academy of Family Physicians (AAFP), is also the President/CEO of DirectTrust.org—a non-profit dedicated to establishing the federated trust framework to facilitate further expansion of Direct use. Dr. Kibbe had this to say about Direct: "The important thing to remember about Direct message exchange is that anyone—yes, anyone—can have an address. Direct is not secure messaging over the Internet based on "membership" in an organization or website. It's really just like e-mail, only with added security afforded by encryption and validation of the senders and receivers and the services acting as their trust agents, as trustworthy. Basically, that means they are who they say they are.

"So, provided the parties trust one another, then providers and patients will be able to communicate with one another on an equal footing, and this gives patients a whole new level of access to their medical information and control over what to do with it.

"But it gets better. Direct message exchange can be used to send messages and attachments to and from devices, for example from home health and well-being monitors of various kinds, securely, at very low cost, and with standard interfacing that will ultimately let many different devices connect on behalf of the patients and the providers who are caring for them."[36]

Some of the initial use cases for Direct include the following:

• Primary care provider refers patient to specialist including summary care record;

• Primary care provider refers patient to hospital including summary care record;

• Specialist sends summary care information back to referring provider;

• Hospital sends discharge information to referring provider;

• Laboratory sends lab results to ordering provider;

• Transaction sender receives delivery receipt;

• Provider sends patient health information to the patient;

• Hospital sends patient health information to the patient;

• Provider sends a clinical summary of an office visit to the patient;

• Hospital sends a clinical summary at discharge to the patient;

• Provider sends reminder for preventive or follow-up care to the patient; and

• Primary care provider sends patient immunization data to public health.[37]

If all goes as anticipated, providers and patients will be working with Health Information Service Providers (HISPs) and other certificate authorities to be issued Direct addresses. In the interim, as is planned in Georgia, some HIEs may list the patient's Direct address alongside other demographic information in a Master Patient Index and forward lab results and other health information to the patient's Direct address, which may be associated with a PHR.

Prescribing Apps

As mentioned in Chapter 8 on mobile devices and in Case Study 23 regarding Michael Zaroukian, MD, PhD, we are at the beginning of the phenomena of providers prescribing apps to patients. Just as Dr. Zaroukian started with one patient, one condition, one idea, and one app, other providers may find once they have looked at a number of apps, recommending one is similar to recommending a book or article.

But what to do when the app is for a specific condition, and the provider has not had the opportunity to "test drive" the app? How do you know it is good?

This is an issue that is currently being discussed. As Chapter 8 states, the U.S. Food and Drug Administration (FDA) may weigh in with some regulations. Meanwhile, Johns Hopkins is researching principles that underlie good health apps, and the company Happtique has established an industry task force to try to develop a framework for vetting health apps.[38] There is a lot of movement in this area, and clarity may be forthcoming. Prior to that time, though, providers will need to investigate apps on their own, and learn from others—including patients—about which apps are helpful.

SCALING IDEAS—PROVIDER LEADERSHIP
Provider Education

To move any of these ideas forward, the technologies and practices need to be diffused as rapidly as possible. We need to shorten the "bench to bedside" diffusion for health IT and specifically for consumer engagement. For the above ideas to scale, provider associations need to begin including consumer engagement as part of continuing education. Some EHR companies are already heavily emphasizing consumer engagement in their user group meetings, which also provide opportunities to learn about best practices and trends. The Healthcare Information and Management Systems Society (HIMSS) and American Health Information Management Association (AHIMA) state and regional chapters are beginning to include consumer engagement in their monthly events, as well as making it a focus at national conferences. HIMSS also offers a number of patient engagement resources through its website and the work of the eConnecting with Consumers Committee, such as the eConnecting with Consumers Toolkit and this book.[39]

Several provider organizations (including the American College of Physicians [ACP], American College of Cardiology [ACC], American Academy of Family Physicians [AAFP], American Medical Association [AMA], and American Osteopathic Association [AOA]), have also collaborated to launch a new initiative "Doctors Helping Doctors Transform Healthcare Through Health IT." Physicians who have already implemented health IT speak to their peers via blogs and videos. Many of these physicians do not describe themselves as "tech-heads," but instead as physicians who are seeking to use the best tools to provide care. Dr. Michael Zaroukian (featured in Case Study 23), a founding board member of this organization, stated: "I am not interested in technology for its own sake. I'm interested in technology only to the extent that it helps me achieve my goals. And my goals are patient care quality, safety and efficiency, decreasing waste and waiting, and improving the overall care and satisfaction of our patients."[40]

As more studies are done regarding patient engagement tools, and their impact on outcomes, we will see them showing up in provider association and other peer-reviewed journals. Providers should watch for these articles, as they can learn from the experiences of other providers who have "walked through the minefield" of patient engagement, learned, and established best practices. As the saying goes, "Wisdom is learning from other people's mistakes."

Willingness to Change/Accept Patients as Partners

Ultimately, the new methods and tools for engaging patients described in this book mean that some prior activities and habits will fall away or be modified—essentially changing clinical practice. We cannot stress strongly enough that it usually takes a physician champion who sees the vision for patient engagement and is willing to lead his or her practice through the transformation. This is analogous to the need for provider champions in EHR adoption, but due to the more personal nature of patient engagement through health IT, it is likely to appeal to a different provider than was the EHR champion.

We should also make clear that though EHR adoption will facilitate use of many of the additional patient engagement tools, it is not necessarily the first step providers can take. For example, text reminder systems, secure email, and prescribing apps can all occur without the use of an EHR.

What is necessary, however, is a change in attitude: from believing that patients should do what they are told to wanting to engage patients as partners in their care. It is an evolution from paternalistic medicine to collaborative or participatory medicine. A good place to start: ask your patients what tools they would like to use to communicate or to self-manage their condition. You may be surprised!

The Science of Personal Transformation/Gamification

Both providers and patients will need to either modify old habits or establish new ones in order to use the tools described in this book. Fortunately, as we seek both to influence patients' adoption of healthy behaviors and help providers and patients to adopt new communications tools, the science of personal transformation can lend us some clues as to methods, even validating engagement in the process.

According to Charles Duhigg, author of *The Power of Habit: Why We Do What We Do in Life and Business*, there "…are now hundreds of researchers, at nearly every major university, studying willpower."[41] We are beginning to understand why some people succeed and others fail, including why some patients tend to do what is good for them. Much of it revolves around the "Habit Loop" identified by researchers at Massachusetts Institute of Technology (MIT): "First, there is a *cue*, a trigger that tells your brain to go into automatic mode and which habit to use. Then there is the *routine*, which can be physical or mental or emotional. Finally, there is a *reward*, which helps your brain figure out if this particular loop is worth remembering for the future."[42]

How is this applied? When a researcher in the UK worked with five dozen orthopedic patients, average age of 68 years old; lower income; recovering from knee and hip replacements, she found that those who wrote out their own specific plans for exercise were "walking almost twice as fast as the ones who had not. They had started getting in and out of their chairs, unassisted, almost three times as fast. They were putting on their shoes, doing the laundry, and making themselves meals quicker than the patients who hadn't scribbled out goals ahead of time."[43]

It turns out that these patients' plans were "...built around inflection points (cues) when they knew their pain—and thus the temptation to quit—would be strongest. The patients were telling themselves how they were going to make it over the hump."[44]

The following four-step framework identified by Duhigg can be used to help both providers and patients to change old routines and establish new ones:

1. Identify the routine (activity to be changed or routine to be replaced: e.g., patient's diet or physical activity; patient's self-monitoring; provider office patient intake form completion; provider use of phone or mail).
2. Experiment with rewards (gamification; lowered insurance premium for patients; reimbursement for providers; satisfaction of "knowing"/relief of anxiety).
3. Isolate the cue (hunger for snack food; need to communicate with provider or patient).
4. Have a plan (patient's wellness plan; provider office's new consumer engagement plan).[45]

The concept of gamification, originally used only for electronic games, is now being applied to motivating patients to use devices or online wellness tools, such as Dr. Joseph Cafazzo's case study (Case Study 5) detailing the use of gamification to increase adolescent diabetics' adherence in taking their blood glucose readings, resulting in a 49.6% increase in adherence.

According to Wikipedia, gamification is defined as "...the use of game design elements, game thinking and game mechanics to enhance non-game contexts. Typically, gamification applies to non-game applications and processes, in order to encourage people to adopt them, or to influence how they are used. Gamification works by making technology more engaging by encouraging users to engage in desired behaviors, by showing a path to mastery and autonomy, by helping to solve problems and not being a distraction, and by taking advantage of humans' psychological predisposition to engage in gaming. The technique can encourage people to perform chores that they ordinarily consider boring, such as completing surveys, shopping, filling out tax forms, or reading websites."[46]

Combined with tools that make changing routines easy and even fun, gamification is poised to make significant inroads in both providers' and patients' use of technology tools in the coming years. Jim Hansen, Executive Director of the self-insured employer consortium Dossia, stated: "Contextual social media, gamification and a substitutable app marketplace played key roles in the evolution of our product from a PHR to the Dossia Personal Health Management System (PHMS)—where engagement comes naturally from bringing one's longitudinal data to life."[47]

Opportunities to Lead/Contribute

There are opportunities for providers to step into leadership roles and help to establish and push standards and the necessary industry frameworks forward. One example of this is participation in membership organizations' committees, task forces, and work groups (e.g., HIMSS, provider associations, or others) to address specific patient engagement issues.

There are also new organizations, such as DirectTrust.org, which is working to establish a federated trust network that will enable the expansion of the Direct secure messaging protocols and address use. It is not a government-funded initiative, but one that will be dependent initially on the energy of volunteers and new members. Another is the Kantara Initiative, which is striving to establish standards for private sector (non-government) authentication of individuals—including patients. Kantara is the only organization that accredits auditing organizations that can visit a hospital or physician office and confirm that their electronic authentication of providers and patients meets a government-established "Level of Authentication" (LOA). Both of these initiatives are striving to ensure that private sector non-profits, in which both providers and patients can participate, are responsible for moving innovation forward in ways that are sustainable and nimble, yet trustworthy.

Legislative, Regulatory, and Government Leadership

Due to the citizen nature of most state legislatures, and the Congress, most federal and state laws do not prohibit provision of information to patients, but some laws may not have thought to include them. As HIEs and state/regional immunization registries seek to provide health information directly to citizens, they are discovering whether any laws need to be changed. Regional provider organizations should look at associated laws and regulations in their jurisdictions and determine if any changes are needed.

An example of a state law with very specific prohibitions or conditions for patient access is the California Health and Safety Code 123148, passed via the Patient Access to Health Records Act in 2002. Though this law facilitates healthcare providers sharing a variety of lab data via "Internet posting or other electronic means," it contains the following clause:

"…none of the following clinical laboratory test results and any other related results shall be conveyed to a patient by Internet posting or other electronic means:
(1) HIV antibody test.
(2) Presence of antigens indicating a hepatitis infection.
(3) Abusing the use of drugs.
(4) Test results related to routinely processed tissues, including skin biopsies, Pap smear tests, products of conception, and bone marrow aspirations for morphological evaluation, if they reveal a malignancy."[48]

This law has raised a great deal of discussion online, and there is a proposed bill to change much of this existing law to one that facilitates greater access, but it still prohibits the automatic distribution of data that might show the conditions specified earlier without a conversation between patient and provider *about each test result*,[49] which means that automatic release protocols, even with time delays, are not an option.

One solution to provider (and legislator) concerns about patient anxiety is suggested by Jason Poston, MD, at the University of Chicago Medical Center (also noted by Ted Eytan, MD, in his blog):

"Dr. Poston, an intensive care specialist, teaches medical students to begin educating patients about results even before the test is done. Patients should have realistic expectations about what results may or may not reveal, he tells students, and why some tests still take time to be analyzed. Not only will patient anxiety be somewhat alleviated, Dr. Poston said, but the role of the doctor as critical guide and partner in the patient's care will be reinforced—even as a patient's need to participate in decision-making will be supported.

"Informed, calmer patients, Dr. Poston added, are more likely to comply with tests, surgery and medication, increasing the likelihood of better outcomes. But a doctor needs time to assess the patient's psychological needs—and doctors can't bill for hand-holding. 'How do we come to a system where we can do right by the patient and also make the practice financially viable?' Dr. Poston said."[50]

Chapter 4 on Patient Safety suggested that immediate patient access to lab results would lower the approximately 7% of significantly abnormal lab results that providers fail to communicate to patients.[51] We would suggest that transparency plus best practices is a better alternative than regulation.

Meanwhile, the federal government, led by the Blue Button Pledge effort, is looking to remove any federal barriers, and the healthcare community is suggesting more patient engagement in future Meaningful Use regulations. Provider organizations may wish to become part of the federal Blue Button Pledge by going to the website and taking the following "Date Holder" pledge:

Data Holders' Pledge

For those who manage or maintain individually identifiable health data (e.g., providers, hospitals, payers, retail pharmacies), we invite you to pledge the following:

We pledge to make it easier for individuals and their caregivers to have secure, timely, and electronic access to their health information. We further encourage individuals to use this information to improve their health and their care.[52]

More information is available on the website regarding the recommended activities for data holders within their practice. Participants can stay aware of opportunities to break down regulatory or legislative barriers through the community.

Certainly we can still expect some more regulatory activity from the FDA in the area of medical devices, which may be linked to apps, or of apps themselves. As discussed earlier, this is an area of great interest and one which the FDA has stated a commitment to try to balance patient safety with facilitating innovation.

CONCLUSION

We have discussed a number of the variables that could influence the future of patient engagement, as well as some of the ideas that appear prepared to scale over the next few years. Though no one can predict the future with precision, we believe that it pays dividends to learn more about these concepts. Large provider organizations may wish to establish a roadmap for implementing patient engagement tools and processes (and may have already done so relative to Meaningful Use). Smaller providers, often more nimble, may have the option of moving faster down some of the aforementioned paths.

We had the opportunity to consult two leaders in both health IT and patient engagement: former ONC Director Robert Kolodner, MD, and Dossia Consortium Executive Director Jim Hansen, and ask their views on the future. The following quote from Jim Hansen incorporates the view of Dr. Kolodner also:

"Beyond the current HIE standards work being conducted today, Former National Coordinator Rob Kolodner MD has it right, 'We are *not yet* at the "model T" stage of health and healthcare informatics. Our current approaches and tools are simply inadequate to support a rapidly transforming health industry.' Myself and others expect that a superset next generation 'intelligent exchange language (or IEL)' will soon be required to handle the incredibly promising yet complex informatics interactions between the three health domains: healthcare delivery, population and personal. Patients who have chronic diseases will be generating huge amounts of data as device-connected 'human body signals.' Genomic sequencing will need to be processed and massive population data stores will need to be queried in order to compare the patient to their virtual current and past cohort. Along with an aggregated longitudinal consumer/patient-centric health record, all of the aforementioned tools will be required to help physicians determine the right personalized diagnosis and treatment pathway *In real time at the 'moment of care' (physical or virtual).*"[53]

Why does the potential future matter for patients?

The seeds of the future are being planted today, in the patient engagement tools and tactics that are beginning to be commonplace. They point to a future in which patients will have the opportunity to have far more knowledge about their health and their healthcare conditions and will be able to act as partners with their providers in getting the best and the most personalized treatments.

Why does the potential future matter for providers?

The pace of change in medicine is accelerating, enabled by new technologies, new policies, and new consumer trends. Many of the changes in patient engagement capabilities have the potential to take the practice of medicine "back to the future," to an era in which providers are supported in practicing high tech *and* high touch medicine. It is all of our responsibility to help this future become a reality.

Why does the potential future matter for the healthcare system as a whole?

Despite the amount of money we spend on healthcare both in the U.S. and around the world, the current system is unsustainable, with misaligned incentives, duplicative and ineffective care, unhappy providers and unhappy patients. Patient engagement has the potential—though not single-handedly—to be a factor that promotes and sustains the necessary changes.

We will see you in the future!

For more information, read the following case studies at the end of this book:

Case Study 4: From the Heart: The Intersection Between Personal Health and Technology

Case Study 5: Social Media, Mobile, and Gamification for Diabetes Care

Case Study 7: Everyone Performs Better When They Are Informed Better

Case Study 8: From God to Guide

Case Study 16: Online Therapy and Patient Engagement

Case Study 20: Good Days Ahead, Online Cognitive Behavioral Therapy

Case Study 22: Veterans Health Administration, Department of Veterans Affairs– Patient Engagement and the Blue Button

Case Study 23: Prescribing Apps for Weight Loss

REFERENCES

1. Fox S. The Social Life of Health Information, 2011. Pew Research Center's Internet and American Life Project. May 12, 2011. Web. Accessed September 9, 2012.

2. U.S. Department of Veterans Affairs. Blue Button reaches one million registered patients. August 31, 2012. http://www.va.gov/opa/pressrel/pressrelease.cfm?id=2378. Accessed September 9, 2012.

3. Cohen SB, Grote KD, et al. Increasing consumerism in healthcare through intelligent information technology. *Am J Manag Care.* December 2010; (16;12 Suppl HIT): SP37-43.

4. 2011 Survey of Health Care Consumers in the United States: Key Findings, Strategic Implications. Deloitte Development, LLC; 2011. http://www.deloitte.com/view/en_US/us/Industries/health-care-providers/center-for-health-solutions/research/index.htm. Accessed September 29, 2012.

5. Ibid.

6. Ibid.

7. Bachman R. Healthcare consumerism: The basis of a 21st century intelligent health system. Center for Health Transformation. 2006. http://www.health carevisions.net/f/2006_Healthcare_Consumerism_CHT.pdf. Accessed September 11, 2012.

8. Diabetes Prevention and Control. Diabetesatwork.org Centers for Disease Control and Prevention. http://www.diabetesatwork.org/GettingStarted/DiabetesControl.cfm. Accessed September 22, 2012.

9. Ten Facts about Chronic Disease. World Health Organization. www.who.int/features/factfiles/chp/01_en.html. Accessed September 29, 2012.

10. Joint Principles of the Patient-Centered Medical Home. Patient-Centered Primary Care Collaborative. February 2007. http://www.pcpcc.net/content/joint-principles-patient-centered-medical-home. Accessed September 29, 2012.

11. An HFM Compendium: Contemplating the ACO opportunity. Healthcare Finance Management Association. December 1, 2010. http://www.hfma.org/Templates/InteriorMaster.aspx?id=24082 . Accessed September 29, 2012.

12. Jones SS, Caloyeras JP, Mattke S. Power to the People: The role of consumer-controlled personal health management systems in the evolution of employer-based health care benefits. RAND Corporation. 2011. http://www.rand.org/content/dam/rand/pubs/occasional_papers/2011/RAND_OP352.pdf. Accessed September 11, 2012.

13. What is Lean? Lean Enterprise Institute. http://www.lean.org/WhatsLean/. Accessed September 12, 2012.

14. Clinician's guide to e-Prescribing: 2011 Update. American Medical Association. 2011. http://www.ama-assn.org/resources/doc/hit/clinicians-guide-erx.pdf. Accessed September 29, 2012.

15. MGMA. Analyzing the cost of administrative complexity – 2004. http://www.mgma.com/about/default.aspx?id=280. Accessed September 12, 2012.

16. U.S. population projections (Internet). Suiteland, MD: U.S. Census Bureau; 2008. http://www.census.gov/population/www/projections/natproj.html. Accessed September 22, 2012.

17. Topol E. *The Creative Destruction of Medicine.* New York: Basic Books; 2012.

18. Davies K. *The $1,000 Genome: The Revolution in DNA Sequencing and the New Era of Personalized Medicine.* New York: Free Press; 2010.

19. Americans support online personal health records; Patient privacy and control over their own Information are crucial to acceptance. Markle Foundation. http://www.markle.org/news-events/media-releases/americans-support-online-personal-health-records-patient-privacy-and-cont. Accessed September 27, 2012.

20. Secretary Leavitt announces new principles, tools to protect privacy, encourage more effective use of patient information to improve care. Department of Health and Human Services. Published December 15, 2008. http://www.hhs.gov/news/press/2008pres/12/20081215a.html. Accessed September 27, 2012.

21. Secretary Sebelius Spotlights New Efforts to Empower Patients to Increase Access to Their Health Information. Department of Health and Human Services. September 12, 2011. http://www.hhs.gov/news/press/2011pres/09/20110912a.html. Accessed September 27, 2012.

22. 2011 Survey of Healthcare Consumers in the United States: Key Findings, Strategic Implications. Deloitte Development, LLC. 2011. http://www.deloitte.com/view/en_US/us/Industries/health-care-providers/center-for-health-solutions/research/index.htm. Accessed September 29, 2012.

23. Ibid.

24. Ibid.

25. Need to Know: Update for Blue Button Partners. U.S. Department of Veteran's Affairs. http://www.va.gov/bluebutton/. Accessed September 29, 2012.

26. Automate Blue Button Charter. S& I Framework Wiki. http://wiki.siframework.org/Automate+Blue+Button+Project+Charter. Accessed September 29, 2012.

27. State Health Information Exchange grantees take part in the consumer innovation challenge. U.S. Department of Health & Human Services. April 11, 2012. http://www.healthit.gov/buzz-blog/state-hie/state-health-information-exchange-grantees-part-consumer-innovation-challenge/. Accessed September 29, 2012.

28. Williams C, Mostashari F, Mertz K, et al. From the Office of the National Coordinator: The strategy for advancing the exchange of health information." *Health Aff.* 2012; 31,3:527-536.

29. Welcome! The Society for Participatory Medicine. http://participatorymedicine.org/. Accessed September 29, 2012.

30. Ed Wagner presented at the IHI National Forum. December 10, 2007. Redesigning chronic illness care: The chronic care model. www.improvingchroniccare.org%2Fdownloads%2Fredesigning_chronic_illness_care__the_ccm.ppt. Accessed September 29, 2012.

31. Healthcare 'friending' social media: What is it, how is it used, and what should I do? Healthcare Information and Management Systems Society (HIMSS). February 12, 2012. www.himss.org/asp/contentredirector.asp?contentid=79496. Accessed September 29, 2012.

32. Counting every moment. Economist.com. May 3, 2012. http://www.economist.com/node/21548493. Accessed September 27, 2012.

33. Ibid.

34. Do allergies cause asthma? Kids Health from Nemours. http://kidshealth.org/parent/medical/allergies/allergies_asthma.html. Accessed September 29, 2012.

35. Forward with Ford 2011: Health and wellness monitoring. Ford.com http://media.ford.com/images/10031/Health_and_Wellness.pdf. Accessed September 29, 2012.

36. Kibbe DC. Personal Interview. September 29, 2012.

37. NHIN direct overview. DirectProject.org Web. http://wiki.directproject.org/file/view/NHINDirectOverview_verl.doc. Accessed September 29, 2012.

38. Jackson S. Johns Hopkins to evaluate efficacy of health apps. FierceMobileHealthcare.com March 19, 2012. http://www.fiercemobilehealth care.com/story/johns-hopkins-test-mobile-medical-apps/2012-03-19. Accessed September 29, 2012.

39. eConnecting with Consumers Toolkit. HIMSS.org. http://www.himss.org/asp/topics_connectingConsumers_toolkit.asp?faid=288&tid=34. Accessed September 29, 2012.

40. Gordon A. Spotlight on Michael Zaroukian, MD: Information that matters in the here and now. Doctors helping doctors transform health care. http://doctorshelpingdoctorstransformhealthcare.org/2011/why-ehr-why-now/. Accessed September 29, 2012.

41. Duhigg C. *The Power of Habit: Why We Do What We Do in Life and in Business.* New York: Random House; 2012.

42. Ibid.

43. Ibid.

44. Ibid.

45. Ibid.

46. Gamification. Wikipedia. http://en.wikipedia.org/wiki/Gamification. Accessed September 30, 2012.

47. Healthcare 'friending' social media: What is it, how is it used, and what should I do? Healthcare Information and Management Systems Society (HIMSS). February 12, 2012. www.himss.org/asp/contentredirector.asp?contentid=79496. Accessed September 30, 2012.

48. California Health and Safety Code 123148. State of California. Published 2002. http://www.leginfo.ca.gov/cgi-bin/waisgate?WAISdocID=05673620488+0+0+0&WAISaction=retrieve. Accessed September 30, 2012.

49. Eytan T. It's been illegal in California since 2002 for a physician to give a woman her pap smear results online. AB-2253 would change that. Ted Eytan, MD Weblog. Published August 1, 2012. http://www.tedeytan.com/2012/08/01/11261. Accessed September 30, 2012.

50. Hoffman J. The anxiety of waiting for test results. *The New York Times.* July 23, 2012. http://well.blogs.nytimes.com/2012/07/23/the-anxiety-of-waiting-for-test-results/?ref=science. Accessed September 30, 2012.

51. Eytan T. Now Reading: Why patients need access to their lab test results—lack of timely follow-up even with an EHR. Ted Eytan, MD, Weblog. July 31, 2012. http://www.tedeytan.com/2012/07/31/11252. Accessed September 30, 2012.

52. Blue Button Pledge: To empower individuals to be partners in their health through health IT. U.S. Department of Health and Human Services. http://www.healthit.gov/pledge/. Accessed September 29, 2012.

53. Kibbe D, Kolodner R, Hansen J. Personal Interview. September 30, 2012.

Chapter 13

Getting Started with Patient Engagement

By Jan Oldenburg, FHIMSS, and Brad Tritle, CIPP

INTRODUCTION

There are many places to start engaging differently with patients, and this book is full of examples. Even as we provide suggestions, next, about how to start, it is important to remember that there are no wrong places to begin; the journey itself will enable change and growth for you and your patients. The steps to begin may be a little different if you are an individual provider rather than if you run or manage a healthcare organization. We have identified key steps to follow in each case.

If You Are a Provider

To move into action, use the checklist in Table 13-1. As you begin, approach all of these options, and others, in a methodical yet optimistic and energetic manner. Introduce your organization to the Institute for Healthcare Improvement (IHI) Model for Improvement that includes the Plan, Do, Study, Act process, if they are not already using it.[1] Do not be stymied by the length of the road ahead of you—what is most important is to take the first step. Many of these actions reflect changes that you can make yourself.

Table 13-1. Steps for Providers for Engaging Differently with Patients

#	Engagement Step	Comments	Y/N
1.	Be an advocate for patient empowerment in your own organization.	Make it clear that you support patient access to records. Share what you learn. Talk to your peers.	
2.	Survey your own patients to understand what they want and need.	How many say they want access to you via secure messaging? How many want to see their records?	

#	Engagement Step	Comments	Y/N
3.	Become aware of your own listening and encouraging behaviors.	Do you encourage patients to view their records? To email you? How much do you listen vs talk during a visit? You are the most important resource for encouraging patients to use online tools and become empowered.	
4.	Provide links to validated medical information online.	If you cannot afford to buy content, link to information sources you consider valid (watch out for ad content). For important information, post your own content or make a video.	
5.	Adopt an EHR that features patient engagement tools or supports APIs to allow you to use third-party tools (e.g., substitutable apps) that will enable you to share clinical information with patients, and provide additional value to your patients and your practice based on the record content.	Many third-party products can support patient engagement if your EHR is not strong in this area. By adopting an EHR that facilitates third-party integrations, not only will you facilitate new and yet unforeseen functionality, but potentially add years to your EHR's lifetime.	
6.	Review and take the Blue Button data holders' pledge and engage with the pledge community at NeCH.	http://www.healthit.gov/pledge/	
7.	Review and take the Society for Participatory Medicine Provider's pledge.	http://seal.participatorymedicine.org/	
8.	Enable patients to reach you via secure messaging.	Consider the HIPAA guidelines and establish and promote capabilities in your practice.	
9.	Begin sharing health information with patients.	Don't know where to start? • Review Meaningful Use requirements on the patient engagement criteria. • Lab results relieve anxiety and get traffic. • Medication lists. • Allergies. • Immunization records. • After-visit summaries.	
10.	Provide online scheduling for patient convenience.	Even appointment request forms add convenience for patients; move to true integrated scheduling as quickly as you can.	
11.	Begin e-prescribing and support prescription refills.	10% more initial prescriptions are picked up when e-prescribing is used; patient convenience in refills means less interruption in therapies.	
12.	Run a pilot using text messages supporting your patients.	You do not have to figure this out alone—there are lots of tools and applications—many of them free—that you can start with. Consider starting with messages supporting patients with chronic illnesses.	
13.	Engage with your patients on a mobile application for health.	Pick something you personally want to work on, pick an application, and engage your patients to work on it with you.	
14.	Experiment with social media.	This book contains many suggestions for getting started with social media—join the conversation. You cannot lead from the sidelines.	
15.	Market the fact that you practice in partnership with patients and provide clinical data to your patients.	Consumers are hungry for providers who practice this kind of medicine; make it clear that you stand for and with patients.	
16.	Always be innovating.	Reading this book was a great start! Keep reading about this topic, asking your patients what they want, trying new things.	

If You Oversee a Healthcare Organization

Your tasks parallel those of an individual physician, but you will need to explore the topic with the awareness that you are acting on behalf of the organization as a whole. We strongly encourage you to start by moving your organization toward Meaningful Use certification, especially Stage 2, as it focuses on many of the core tasks for patient empowerment. As you begin, approach the work in a methodical yet optimistic and energetic manner. Use the IHI Model for Improvement that includes the Plan, Do, Study, Act process, if your organization is not already using it.[2] Do not be stymied by the length of the journey—what is most important is to take the first step. To move into action, use the checklist in Table 13-2.

Table 13-2. Steps for Provider Organizations to Facilitate Provider Engagement with Patients

#	Engagement Step	Comments	Y/N
1.	Offer training for providers on engaging patients; teach them techniques for listening and motivational interviewing; consider developing relevant continuing medical education (CME) content.	Many providers are unaware of how little they listen and how dismissive and discouraging their behaviors may seem to patients.	
2.	Build organizational support for patient-centered care.	Survey patient satisfaction regularly; reward providers who get high marks with compensation. Show providers their rankings against others.	
3.	Search out provider champions.	Identify providers in the organization who already practice participatory medicine and who have the respect of their peers. Support them in spreading the word and leading by example.	
4.	Establish policies to support online capabilities for patients.	A policy framework can support and enable innovation, while providing guardrails for it.	
5.	Build a library of educational tools or links for patients as part of your enterprise website.	You do not have to create content, nor purchase it. You can direct patients to existing, validated medical content and selectively choose to enhance that content with high-value information your organization creates.	
6.	Implement tools and a framework to support secure messaging between doctors and patients.	As you implement, provide training on how to use secure messaging effectively; consider creating triage relationships within email to support a care team approach.	
7.	Adopt an EHR that features patient engagement tools or supports APIs to allow you to use third-party tools (e.g., substitutable apps) that will enable you to share medical records with patients, and provide additional value to your patients and your practice based on the record content.	Many third-party products can support patient engagement if your EHR is not strong in this area. By adopting an EHR that facilitates third-party integrations, not only will you facilitate new and yet unforeseen functionality, but potentially add years to your EHR's lifetime.	
8.	Begin sharing clinical information with patients.	Don't know where to start? • Lab results relieve anxiety and get traffic. • Medication lists. • Allergies. • Immunization records. • After-visit summaries.	
9.	Pursue Meaningful Use Stage 1 and 2 certification.	Stage 2, in particular, supports key patient engagement goals.	

10.	Review and take the Blue Button data holders' pledge on behalf of your organization and join the pledge community.	http://www.healthit.gov/pledge/	
11.	Encourage physicians in your organization to take the Society for Participatory Medicine Providers' pledge. Drive for all of your providers to "get the seal."	http://seal.participatorymedicine.org/	
12.	Provide online scheduling for patient convenience.	Making it convenient to interact with your organization tells patients that you care about their experience and their time.	
13.	Begin e-prescribing and support prescription refills.	This can save your organization time while increasing patient engagement.	
14.	Create a culture of innovation. Encourage providers to experiment with new tools and ways of interacting with patients.	Many of the examples in this book highlight innovation opportunities around patient engagement.	
15.	Market patient engagement as one of your organization's strengths.	Make sure you have the capabilities in place and your staff behind you when you do this, so that your brand promise matches patients' actual experiences.	

For further information, please explore the HIMSS website: a HIMSS eConnecting with Consumers toolkit is updated regularly. It can be found at: http://www.himss.org/ASP/topics_connectingConsumers_toolkit.asp?faid=288&tid=34.

CONCLUSION

Several providers and organizations we interviewed voiced disappointment at their patients' adoption rate for these tools. You can help make adoption a reality by making sure everyone in the office talks about the capabilities, that you have staff available to help your patients get registered, and that you remind patients at every turn that they have online ways to access their own information, as well as direct ways to contact you. Most patients do not interact with the healthcare system daily, the way they shop or bank. It may take some repetition for patients to think first of their online options when they need healthcare. Over time, however, your patients will warm to the convenience and online interactions will become habitual for them.

In the process, your patients will become more health literate, more empowered, more self-actualized: not only knowing the actions they can take to improve their health, but also acting appropriately. By helping your patients to feel more responsible for their health, and providing them with information and tools that fuel this empowerment, we all win. Engaging patients—and having engaged patients—is not a destination—it is a journey. Having learned through this book from those further down the path, let us join hands, and take the first step.

REFERENCES

1. Institute for Healthcare Improvement. August 28, 2012. http://www.ihi.org/knowledge/Pages/HowtoImprove/default.aspx. Accessed October 10, 2012.

2. Institute for Healthcare Improvement. August 28, 2012. http://www.ihi.org/knowledge/Pages/HowtoImprove/default.aspx. Accessed October 10, 2012.

Case Study 1

Patient Engagement in a Patient-Centered Medical Home

By Jan Oldenburg, FHIMSS (From an interview with Ken Adler, MD, MMM)

Dr. Ken Adler and his three colleagues have been recognized by the National Committee for Quality Assurance (NCQA) as a Patient-Centered Medical Home (PCMH) practice for over three years. They are part of a larger medical group with 122 doctors in 50 clinical offices. Dr. Adler's practice has been using a set of tools to enable secure e-messaging with patients for nearly five years; at this time, the entire medical group is implementing a full patient portal that they expect to be operational in all 50 practices within the next six months.

A long-time advocate for e-prescribing and providing patients with access to their records, Dr. Adler knows that engaged, activated patients are the cornerstone of an effective PCMH model. Using secure messaging as one tool available for engaging patients, he responds directly to patients, without any office intermediary on certain clinical issues.

Dr. Adler said, "My only disappointment with patient engagement has been in the low adoption of secure e-messaging by our patients. In a pre-implementation survey, more than 80% said that they wanted it, but less than 10% have signed up. I believe that the addition of access to their medical records will dramatically improve that adoption. I know we could force adoption if we told patients that we would no longer snail-mail them their lab results but would instead only discuss them in person or send them electronically. That is likely to be the trend in the future—for us and others."

He generally encourages patients to refill their prescriptions through community pharmacies because that way his office does not need to touch the prescription unless there are no refills left. Refills that need authorizations can be sent from the pharmacy to the clinic electronically and processed as a part of his workflow.

In the PCMH pilot, Dr. Adler is paid three ways: fee for service, a monthly capitation fee, and a bonus based on meeting quality metrics. Practicing in this way has made him aware of how inefficient it is to bring patients into the office for all care, but he notes that there are no clear metrics yet about how to "right size" care. It would help providers move from fee-for-service medicine to population health if there were more standards of practice addressing when to bring someone into the office versus offer virtual care.

Despite some of the constraints presented by the tools available, Dr. Adler believes that when patient portals are done right, they can make a practice more efficient while also satisfying patients. He is looking forward to implementing even more sophisticated patient engagement tools that will result in increased satisfaction for both patients and providers.

Case Study 2

Innovation in the Context of Meaningful Use

By Jan Oldenburg, FHIMSS (From an Interview with Jen Brull, MD, FAAFP)

Dr. Jen Brull was the first person in Kansas and the 34th in the nation to attest to Meaningful Use Stage 1, partly because of her love for health information technology (health IT) and health information exchange (HIE). She began working on Meaningful Use the first day Stage 1 criteria were published. Her software vendor was not yet ready for certification, but she suggested a grand bargain: she would help the vendor get to Stage 1 certification by testing the features; in return, her vendor provided access to the patient portal free—and installed the Stage 1 version of software Christmas week of 2010 so that she could meet her goal of attesting early in 2011. For Dr. Brull and her clinic, the whole Meaningful Use attestation experience has been positive, despite the hard work to achieve it.

Dr. Brull lives and works in a small community, where she connects with her patients on many levels—as patients, yes, but not just as patients. They are also the parents of her children's friends, the people whose shops she frequents, the people she meets at the movies or in church. Those many-layered connections inevitably make for nuanced, multi-faceted interactions, where a friend she is linked to on Facebook may reach out with a medical question or someone she knows from her children's school may send her an email. An example of the way this works in a small town was the family from Dr. Brull's practice who moved to Kansas City, several hours away. When their child had a sore throat, the parents took her in to an urgent care clinic, where the doctor found that the child's blood sugar was incredibly high. The urgent care doctor told them to go to the emergency department (ED), but in a new city, and without insurance, they were worried about the cost and unsure what to do. The mother sent a message via Facebook to Dr. Brull. Dr. Brull called her, and was able to arrange for the daughter to be seen at a friend's clinic at a lower cost than an ED visit. These kinds of connections simply would not be logical or practical in a larger community.

Despite Dr. Brull's deep belief in both patient empowerment and personal health IT, it is doubtful that her patients would have a portal without Meaningful Use, primarily because of the cost of the initial investment. The arrangement she made with her vendor around testing Meaningful Use Stage 1 in exchange for free access to the patient portal made it reasonable for her to make portal capabilities available to her patients in Stage 1.

Registering patients on the portal currently requires collection of an email address and signing of a consent form. Many patients are reluctant to provide their email address without seeing the value up front, so only 14% of her patients are currently registered. Dr. Brull is working with the vendor to make registration more straightforward, so that patients can see the portal's value before providing an email address for contact. This change will also be required to enable the practice to demonstrate that they have met Meaningful Use Stage 2 access requirements.

Dr. Brull explained some of the benefits of the patient portal as follows:

• Some patients require a disproportionate amount of time to collect their thoughts and express themselves when you call them on the phone. The asynchronous nature of email means they can take all the time they need to compose their email, and the doctor can respond very efficiently.

• Patients love the ability to communicate directly with their physician, and especially like the fact that they can send an email when they are anxious—even at 2 am—and get a response the next day.

• Many patients in Dr. Brull's small town in Kansas travel—a lot—some for business, some for pleasure. Email contact makes it really easy for them to address a problem while they are on the road. With e-prescribing, it is easy for Dr. Brull to find a pharmacy anywhere in the country to send a prescription in a location where her patient can easily pick it up while on the road.

Today's patient portal is functional but not robust; there are a host of improvements that would make it far better. She likens today's patient portals to early mobile phones that weighed over two pounds, were nine inches long, and needed to be charged for ten hours to enable users to talk for 35 minutes. Inevitably, market forces and innovation will continue to enhance the way patients and providers communicate and exchange data.

While nothing in Stage 2 Meaningful Use looks particularly difficult to Dr. Brull, if she could wave a magic wand to get everything she needs, she would wish for dramatic improvements in health information exchange. It should be ubiquitous, secure, and should make it simple to find patient histories, wherever patients were previously seen. She gave the example of a cardiologist who is seeing one of her patients and discovered that the patient had an electrophysiology (EP) study done ten years ago at a Kansas hospital. He asked for Dr. Brull's help in finding the hospital where the procedure was performed so he could get a copy of the historical study and compare it to a current one—no easy task today. What if, instead, it was as simple to find a record like this by logging onto a secure portal, establishing their credentials as doctors and their reason for seeing this patient, and then "Googling" the patient's data? Yes, they would still need tools to normalize and make sense of the data, but the potential to make a real difference in patient care by simplifying the exchange of data is enormous.

Case Study 3

How Social Media Has Changed My Medical Practice*

By Natasha Burgert, MD, FAAP

Last summer, I joined millions of others in the deluge of social media. I committed one year of effort to see if social media would enhance or distract from my pediatric practice.

That was my goal, just one year.

At that time, I wanted to dip my foot in the pool and see if it made any ripples. The unexpected consequence was how much social media has changed my medical practice, and me. Ripples have returned as tidal waves.

My practice has seen tangible, real valuable benefits. I have been intellectually challenged and have professionally grown.

For my practice:

- Increasing new patient traffic is creating revenue for our group. I average one new patient family per week who came because of our social media presence. I know this because they tell me, "I am here to see you today because I found you on Facebook," or "I found your blog."

- Fifty-two patients a year x $2,700 (average pediatric care for 0-24 mon.) = $140,000 of average billable income over two years.

- Creating information has added to my "searchability" in search engines. All my work is available publicly and with fully disclosed authorship, so new patients can find me with ease.

- Investing time in relevant and complete posts actually saves me time in the long run. Questions I am repeatedly asked, like "How do I start solid foods?" can be answered quickly and completely by directing them to my site. This saves face-to-face clinic time for more specific concerns for their child.

- I have created opportunities to make my families' lives easier by using the technology at their fingertips.

- Selectively following leaders in the field of pediatrics has allowed me to refresh and update my knowledge daily. The lead article in medical journals, the newest recall, the updated reports are in my information stream. Sharing the headlines

---

*From a posting at KevinMD.com, August 18, 2011. Used by permission.

and reports that will most assist my patients continues the information stream in real time.

- I can get help for my patients across the country through online professional connections, and I have experts at my fingertips who can help me answer questions.

For me:

- Being part of the health social media and blogging community has given me a connection and an outlet. I can express myself as a physician and a mom, creating a "professional diary" of my life.
- I have met amazing people with big ideas and bigger hearts, who inspire and challenge me daily.
- I have seen a glimpse of how big an effect a group of vocal health writers can have; how active advocates can act to correct falsehoods and incorrect reporting. I am a part of a movement; a way that healthcare is changing.
- I unexpectedly found how one's purpose could be defined, in such a short amount of time.

For my patient families:

- I can actively communicate, acknowledge, and positively influence the choices that my families make for their children between checkups. My anticipatory guidance can be repeated, reinforced, and repeated again.
- New websites, blogs, and apps are constantly being added to our fingertips. After review, I can refer my patients to some really cool, applicable technology options to better care for their kids. I would never know about this stuff if I was not involved with social media.
- I can act as a "filter" to promote the good and refute the bad.
- I can be a source of reliable, real information.

But what is all of this really about?

- It's about the mom who comes to me at the 18-month check up and tells me her child's car seat is still rear-facing.
- It's about the dad who tells me he went to the health department and got a TDaP (tetanus, diphtheria and acellular pertussis) vaccine before his new son was born.
- It's about the complete stranger who sees me in my office building and says, "Are you Dr. Natasha? Thanks for writing about kids and fever. I had some questions, and it came at just the right time."
- The beauty of social media is that I never talked with these parents about these health and safety issues. Parents made good decisions for their families after getting the information. Period. That's all they needed, and that's all it took.

     Wow.

     Offering online authenticity, genuine concern, and experience (sprinkled with a bit of sound medical knowledge) has created an amazingly powerful platform, and helpful practice tool.

Although using social media does have some undefined, grey areas to navigate, for me one thing is clear: my goal of one year has been extended until further notice.

Case Study 4

From the Heart:
The Intersection Between Personal
Health and Technology

By Nancy Burghart-Hall, CHCIO

My name is Nancy Hall. I am a healthcare IT executive, and have worked in the field for more than 20 years. While my professional life story has focused on healthcare IT, like many of you, I also have a personal health story.

We live our lives not thinking about cancer, chronic illness, disease, death—these things happen to other people.

In 1998, my husband, Tom Hall, was diagnosed with multiple myeloma, a rare and terminal blood cancer. He was 48. His life was in full swing with a family, a thriving business. There wasn't time for slowing down or getting sick, but cancer does not discriminate.

As Tom's wife, I took on my new challenge with confidence. After all, I worked in healthcare at one of the largest and strongest health systems in the country. Our cancer center was a nationally recognized center of excellence; I knew the system, the doctors, the administrators. I settled into my new role—caring for a loved one with cancer. I learned everything I could about multiple myeloma, chemotherapy, and radiation, exploring the latest research to figure out what our options were. Our oncologist was brilliant, supportive, and open. He listened and cared; he included us in all discussions of treatment options. Our hundreds of interactions were always collaborative, and we developed a trust that is unique and treasured.

The best doctor. The best health system. We were fortunate. I believed we were in control of our situation—as much as anyone can be in control while fighting a terminal illness.

But being in control without having information is really not possible.

Life is precious, especially when it's short, and for brief moments, our lives seemed almost normal.

For three years we stayed ahead of the cancer. We were outrunning it, outwitting it, enduring monthly treatments, radiation, chemotherapy, and even a stem-cell transplant. One evening, during a brief moment of calm, we were at home watching a movie. Mel Gibson was fighting for the freedom of all Scotland, and things were going okay for us too. Toward the end of the movie, unable to watch the final torture scenes, I left the room. Less than 15 minutes later, I heard muffled sounds and then a large thud. I ran down the hallway, only to find my husband collapsed on the floor, unconscious and unresponsive.

I dialed 911, explained what had happened, and asked for an ambulance. The 911 operator asked me about Tom's medical history, then told me to "gather all his medications," and wait for the paramedics. Reluctantly I left my husband on the floor. Alone. Probably dying.

While I was searching the medicine cabinet, panicked, for Tom's medications, I heard paramedics come into the house. As I was hurrying back with the medications, I heard one of them say, "Slow down. I think he's a DNR." Despite my career in healthcare, I'd never heard that acronym before. "DNR?" I asked. "Why aren't you saving his life? Help him!!" I shouted. "He's not a Do Not Resuscitate! He's not done yet!"

I grabbed the paramedic's arm. I told them we were watching TV, that he was fine 15 minutes ago. I wondered why I had to convince them he was worth saving. Yes, he had cancer. Yes, we were probably losing the battle. But these were precious moments in our life, and they were wasting precious seconds debating whether he was worth saving. The nearest hospital was only two blocks away, and I knew we could get to the ER fast—and we did, when they finally took him. But this was the "competing" health system. We were patients at the health system further away, where Tom's medical records, previous ER visits, and chemotherapy treatments were stored, and where his doctors practiced. Because Tom was critical, the paramedics took him to the closest ER, where we weren't insured, no one knew us, his medical history wasn't available, and our doctor didn't have admitting privileges.

When we arrived, the triage nurse took me aside and asked me his medical history, whether I'd brought his medications, whether I could explain what happened. I recounted the story—for the third time that night. It seemed like hours went by, and while I was talking with the nurse, they intubated and tried to stabilize my husband. I still didn't understand what could have happened in 15 minutes that had changed everything for us.

It gradually became clear that he had gone into septic shock and that 40% of septic shock patients die, even with treatment. It was beyond my comprehension. He had survived this far with terminal cancer. I understood that, but I didn't know what to do with whatever was happening now. At some point during that long night, I was asked what his wishes were regarding end-of-life efforts and encouraged to notify family members.

I wondered what happened to our oncologist, Tom's doctor for the past three years, who had all the knowledge and insight on his case. He'd saved Tom's life this far, but neither he nor the critical information he had about Tom were with us in the ER. He was affiliated with the competition and could not see patients in this hospital.

I was exhausted, confused, and alone.

Thankfully, my husband did not die that night. They moved him to the ICU, where he remained for five days, then to an oncology floor for another two days before he was stable enough to be transferred to a hospital in our health system. During that week, his doctor could not treat him and his medical history was not available. I knew they were performing unnecessary tests; tests he'd recently had in our own health system, but they were not available to the doctors seeing him now. It was like being in another country—but we were seven miles away in the same town.

I don't know how many times that week I was asked about his medical history, when all I wanted was to sit by his side, hold his hand, tell him, "We'll get through this one." Mostly, though, that wasn't my role. Instead, I answered questions and repeated his history each time someone new came into his room or the shifts changed.

I, primary caregiver, spouse, and chief griever, was also responsible for Tom's medical history, health information exchange, and coordination of care—without a medical record, from memory. A memory that was under duress, exhausted and burdened.

What If...

Our experience that night doesn't have to be the norm. If we are successful in accomplishing the goals of healthcare reform and the HITECH Act, much of the unnecessary pain of that night will be eliminated for others.

Imagine that the following things were true about our experience: My husband and I each have a personal health record (PHR). All the health systems and hospitals in our town have an electronic health record (EHR). Our doctor's office also has an EHR. Our region has a health information exchange (HIE) that connects all the doctors and hospitals, as well as the pharmacies, labs, radiology and oncology centers in our region. Disparate and competing health systems share information and collaborate on patient care through the HIE. When my husband becomes ill, I am able to connect online with national and local caregiver groups for caregivers of cancer patients, specifically those suffering from multiple myeloma. We are connected to a "digital library" of information, informed of new treatments, research, clinical trials, and successes happening with multiple myeloma treatments throughout the world. Other patients tell us about some of the possible risks that accompany treatment for multiple myeloma, including septic shock. We store critical information in the cloud, so that in an emergency, we can access it from a mobile device, anywhere.

In the new world of healthcare enabled by information technology just described, here's the story I might be telling about that night:

One evening while we are watching *BraveHeart*, my husband collapses on the floor, unresponsive. I call 911, and when the operator asks me the situation, I tell him my husband is a cancer patient, possibly in ST (septic shock)— because I know the risk factors for and symptoms of septic shock. I tell the 911 operator that we are connected electronically with our doctors and health system. The ambulance is dispatched, and I never leave my husband's side.

The arriving paramedic uses his iPhone app to digitally scan the 2D barcode on the wall by the front door. Instantaneously he has access to my husband's critical information including his advance directive, and knows that he is not a DNR.

The paramedics immediately begin work saving his life. There's no second guessing, no questions, no arguments. The paramedic's record is digitally transmitted to the hospital in preparation for receiving the patient. The ED doctors have full access to my husband's emergency record and know his condition, medications, and current treatments. Telemedicine is available from the ER to the ambulance so they can monitor and guide the paramedics while in transport.

Our oncologist is notified electronically that his patient is in transport to the ED. Our health system is notified that their patient is in transport to the competing health system's ED. In the ED, unnecessary tests are avoided, and the doctors have access to his recent test results. The doctors in the ED have enough information to determine best treatment paths and options and are able to consult with my husband's oncologist remotely.

Most importantly, I am not asked for Tom's medical history and current medications. Instead, I am comforted and someone asks if they can get me coffee while I sit by his side.

My husband is stabilized and transferred to the ICU. The attending doctors have access to all his medical records, medication regimen, comparison tests, what his "normal" is—there is no guessing. The physicians collaborate with our oncologist. Tom is stabilized faster and moved to our hospital in three days rather than a week. I stay by my husband's side throughout, able to focus on supporting him.

Why Does This Story Matter?

We survived this incident, and Tom went on to live another six months: a lifetime, when you're counting seconds.

During times of great stress, as humans we somehow summon up the strength to "get through it," not even recognizing where the strength comes from. We summon the courage and the strength to go forward, hoping that we will be able to look back and process the experience later from a place of safety. Inevitably, though, when we have time and safety to process the experience, some of the details have been lost.

Medical records are a detailed accounting of a very specific and important part of this journey. They recount health and illness, the actions taken, and the reasons for them. They document a part of who we are. I have no medical records from Tom's journey to help me remember and process the experience, no medical records to help me tell my children the story of what happened that night or during the rest of my journey. And I didn't have medical records available to share with emergency medical technicians and doctors when it mattered the most, when every second counted and accuracy was critical.

But this doesn't have to happen to you. I want to encourage you to participate in healthcare technology on a personal level. If my story—our journey—has inspired you in any way, and you are wondering what your next step might be, become an advocate for healthcare technology by actively participating in and managing your personal health information. You can start right now; you can start for free.

One example of a free PHR is Microsoft HealthVault, which can be located at http://www.microsoft.com/en-us/healthvault/. There are others, too.

Ask your doctor, pharmacist, lab technician, and dentist for your information. Start your journey of tracking your health information and the information of your loved ones. Help us carry this message beyond the medical community.

Support healthcare reform, which will help promote the systemic changes that are needed to make the rest of my story a reality.

And treasure each moment and each day you have with your loved ones.

In Loving Memory,
Tom Hall,
1949–2001.

Case Study 5

Social Media, Mobile and Gamification for Diabetes Care

By Brad Tritle, CIPP, and Karen Colorafi, RN, MBA, BScN, CPEHR, CPHIT (From an Interview with Joseph Cafazzo, PhD, PEng)

Joseph Cafazzo is Centre Lead and Senior Director of Medical Engineering and Healthcare Human Factors at University Health Network (UHN) in Toronto. As director, he leads applied research in the areas of extending the use of medical technology from the hospital to the home and use of human factors methods for the design and evaluation of medical technology. His recent work has been in the area of empowering patients and providers with technology that facilitates self-care. Dr. Cafazzo has published extensively on the use of patient-centered technologies that improve health outcomes and reduce burden on the healthcare system, notably in adults with congestive heart failure (CHF), hypertension (HTN), and children with diabetes.

Officials at the Ministry of Health in Ontario, Canada, were struggling to understand why rehospitalization rates were so high among patients with chronic disease, given the plethora of remote monitoring devices available. Dr. Cafazzo and his team explained that existing technology was expensive, worked by plugging into land lines, and was clunky and difficult for older adults to use. He explained, "We found a total lack of innovation in the remote monitoring industry, even though these were high-margin products."

In part as a result of that conversation, Dr. Cafazzo and his team began working on a project that uses cellular phones, arguably the most ubiquitous product on the market, as a platform for remote patient monitoring. This simple technology has penetration in every age group, from teenagers to the elderly. Their product, BANT, named after Canadian insulin developer Dr. Frederick Banting, uses principles of gamification to encourage children at the Hospital for Sick Children in Toronto, Ontario, to measure and record their blood glucose at various times throughout the day. By doing so, the children earn points that can be redeemed via iTunes for music or games. The pilot, which boasts a 49% increase in adherence and an 87% satisfaction rate, taught the group some important lessons. Dr. Cafazzo estimates that the improvements they observed were likely a combination of the gaming principles and the social-media qualities of the app. "The kids were highly competitive," he notes. "They would discuss how many points they had within the group and even chose to share actual blood sugar values with each other." The team also learned that self-management tools have to be easy. Kids with insulin pumps, for example, had to carry a second meter which made self-reporting more cumbersome and less likely.

The most important finding from a systems perspective was that health outcomes improved, as measured by the HgA$_1$C, but the burden on the primary care provider (PCP) did not. Dr. Cafazzo explained, "All of this is a means to an end. Healthcare spending is out of control in both of our countries. If we can encourage patients to self-manage and lower their dependence on the system through mobile apps and remote monitoring, we've done a great service to the system without increasing the burden on our PCPs."

Any provider can direct their patients to the iTunes store for a version of BANT. Because of regulatory restrictions, it is a simplified version that allows for manual data entry only. They are currently seeking Canadian and American FDA (U.S. Food and Drug Administration) approval, and hope to release a version with automated entry by early 2013.

Case Study 6

Kaiser Permanente–
My Health Manager

By Kate T. Christensen, MD

Kaiser Permanente is one of the United States' largest integrated healthcare organizations, providing both insurance coverage and care to more than 9 million members in nine states and the District of Columbia. The healthcare organization has implemented an electronic health record (EHR) system, Kaiser Permanente HealthConnect®, across all of its clinics and hospitals and offers online record access and transactions for members to help them manage their own health.

Kaiser Permanente's personal health record (PHR), My Health Manager on kp.org, is directly connected to KP HealthConnect. Currently, My Health Manager has more than 4 million registered members, 64% of all eligible members. Usage of the system is high, as shown by the 2011 statistics below:

- 104 million visits to kp.org
- 74 million sign-ons to My Health Manager
- 29.7 million test results viewed (most are immediately available after being reported to the chart whether normal or abnormal)
- 12.2 million email messages sent through a secure messaging system
- 10 million prescriptions refilled (30% of all prescription refills)
- 2.7 million appointments booked

In addition to the features just noted, users of the system are able to access their medical records, including medications, allergies, immunizations, ongoing health conditions, and past visit information. Members can also sign up to receive healthcare reminders and estimate the cost of care. They can also access My Health Manager on behalf of their children or another adult, under a carefully designed proxy authentication program.

Here is how some users describe the capabilities:

- "I am a very active and healthy person, but having a chronic condition like diabetes requires that I take extra care in managing my health," said Kaiser Permanente member, Mary Levine, 59, of Oakland, California. "Having access to the online resources and my physician through kp.org makes me feel confident that I have the tools I need to better manage my health and well-being."
- According to Kaiser Permanente member Kristen Batten, of Alameda, California, having such quick and easy access to her providers via email enriches the

doctor-patient relationship, resulting in increased loyalty to Kaiser Permanente. "I have a connection with my doctor that I've never been able to have before," said Batten. "I feel like at any given moment, she's there for me."

Kaiser Permanente has several published studies about My Health Manager usage and impact. Highlights of their studies are noted:

- EHR use leads to 7% to 10% fewer office visits and 14% fewer telephone contacts (*Health Affairs*, Vol. 28, 2009).

- Online behavior change program use leads to increased productivity at work (HealthMedia, 2008).

- Patients booking online appointments are more likely to keep them (KP internal study, not published, 2008).

- Secure patient-physician email messaging improves the effectiveness of care for chronically ill patients. Specific improvements associated with use of secure messaging with physicians included:

    o Diabetes control (4.9 percentage points) and screening (4.8 percentage points)

    o Cholesterol control (6.5 percentage points) and screening (5.3 percentage points)

    o Blood pressure control (3.2 percentage points)
    (*Health Affairs*, 29, NO. 7, 2010)

- Patients with online access to My Health Manager are 2.6 times more likely than nonusers to remain Kaiser Permanente members (*American Journal of Managed Care*, July 2012).

Kaiser Permanente expanded the access of My Health Manager to mobile devices in early 2012. By summer 2012, 17% of traffic to kp.org came from mobile devices.

There are several key lessons to be learned from the Kaiser Permanente experience.

- Members are keenly motivated to access some parts of a PHR, in particular test results, secure email, refills, and appointments.

- It is essential to make it as easy as possible to register and sign on while maintaining high security standards. At Kaiser Permanente, members can register for the service online, and a security system similar to that used by many banks enables one-step authentication. Introduction of this one-step process made a significant difference in the number of members who completed registration.

- Kaiser Permanente's integrated system allows it to reinforce "sign on to kp.org" messages at touch points throughout the system: by staff at clinic visits, in the lab when blood is drawn, in hospitals, on bills, on claim forms, on enrollment materials. No single marketing effort was the key—instead, it seemed to be the combination of efforts across many venues.

- Aligning incentives for all users and stakeholders is essential. Because Kaiser Permanente is predominantly a staff model organization, where physicians are paid on salary rather than on a fee-for-service basis, the organization overall benefits from reductions in visits due to email or lab results viewed online.

Case Study 7

Everyone Performs Better When They Are Informed Better

By Jan Oldenburg, FHIMSS (From an Interview with Dave deBronkart)

Dave deBronkart is the improbable hero of an improbable tale. His story both mirrors the trajectory of the participatory medicine movement and has helped create that same trajectory. He is best viewed, perhaps, as embodying the Zeitgeist of the 21st Century movement toward patient engagement and patient empowerment.

This part of Dave's story begins in January 2007. Dave, newly returned to the Boston area and a self-described marketing data geek, had a shoulder x-ray to explore some pain he was having. His doctor called him afterwards to say, "Dave, your shoulder's going to be fine, but there's something in your lung. I've ordered a CT for you." Days later, he got the bad news. He had kidney cancer, which had metastasized to his lungs; they would later learn that it had also spread to his leg, arm, skull, and eventually, even his tongue. The best estimate of his median survival was 24 weeks; websites he consulted described his prognosis as "poor," "bleak," and "grim."

Dave was getting his care at Beth Israel Deaconess. He received exceptional care and was able to partner with his doctors at every turn along the way. His primary physician, Danny Sands, MD, suggested to him shortly after his diagnosis, "Dave, you're kind of an online guy—you might like this online community, Association of Cancer Online Resources (ACOR)." Dave was already reading everything he could about his condition and eagerly went to the ACOR site and posted a message to learn more from his peers, just as he had done for years in other areas of life. Within two hours, he got responses from other patients. They suggested that he should ask about a drug called High-Dose Interleukin-2 (HDIL-2). They told him only a small number of patients were helped by it, but that sometimes it produced a complete and permanent response, yet often patients did not even hear about it as an option.

Beth Israel Deaconess was using it, however, and was already planning to find out if Dave could qualify medically. Dave had two primary tumors, which had already erupted beyond the kidney. The kidney was removed in March, and he had his first dose of HDIL-2 therapy in April. Tests in June showed that he was responding, so he had another round in July. In September, his doctors told him he appeared to have beaten "this thing." Two years later he was able to dance at his daughter's wedding, and he is still going strong.

During November of 2007, Dave began to blog about the "New Life of Patient Dave." In January of 2008, a year after his diagnosis, Dr. Sands invited him to a retreat with other friends of their mentor, Tom Ferguson, MD. Dr. Ferguson had been writing a white paper for the Robert Wood Johnson Foundation, entitled "E-patients, How They Can Help Us Heal Healthcare," when he died in 2006. The paper was finished by colleagues. It was Dr. Ferguson who coined the term "e-patients" to describe online individuals who are equipped, enabled, empowered and engaged. Dave saw himself in the paper's stories of patients learning about their conditions in conjunction with and sometimes in parallel to their doctors. Within days he changed the name of his blog to "The New Life of *e-Patient* Dave" and at the retreat found a community of like-minded futurists. Writing and researching about online health became a hobby and soon, a vocation, for Dave.

Early in 2009, Beth Israel Deaconess became one of the first organizations to allow patients to move their clinical records to Google Health, and e-Patient Dave decided to become one of the first patients to take advantage of the opportunity. His motivation was to help kindle a new ecosystem of health data that he believed could not happen without a pool of data for innovators to use. From his high-tech career, he knew that to innovators, data is fuel: no data, no invention. Much to Dave's surprise, when he reviewed the data that had been moved to Google, he found garbage—obsolete information and conditions he had never had. Ultimately, with the help of Ferguson's colleagues, he realized it was not his clinical data that had been sent, but his billing data. He found the record littered with examples of things he had been tested for but found not to have, such as metastases to the brain or spine. They had been billed correctly but, when used as a proxy for clinical data, created the wrong impression. Dave blogged about the experience, of course, and much to his surprise, the story was picked up by *The Boston Globe*. Not just picked up, in fact, but placed on the front page on April 13, 2009.

Within two months, Dave was testifying in front of the National eHealth Collaborative (NeHC) about the importance of enabling patients to access their own medical data, verify its accuracy, and carry it with them to another provider. Nine of the 12 NeHC commissioners who listened to e-Patient Dave at that committee meeting also served on committees defining Meaningful Use. Patient engagement was absent from the first draft of the Meaningful Use provisions that were issued before Dave's testimony; they were present in the second draft after his testimony. E-Patient Dave's timely post affected the trajectory of patient engagement in the U.S. by drawing attention to the rights of patients—and the reality that patients can understand their health records—as well as the problem of providing them with accurate as well as timely data, and the need for structured data from EHRs. Early in 2010, he was invited to Washington, DC to represent the patient viewpoint at two workgroup meetings in the Meaningful Use development process.

These experiences marked the start of e-Patient Dave's switch to a consulting career as a writer and international speaker about patient engagement, patient empowerment, and the power of collaboration between patients and providers. I was privileged to be in the audience for a presentation that e-Patient Dave did in December, 2009 in San Francisco. He is a compelling speaker. As he recounted his story in the front of a room holding nearly 125 people, the group was so still and so mesmerized that motion detectors turned out the lights in half of the room.

When Dave talks about the remarkable trajectory of the participatory medicine movement in gaining acceptance from the establishment, he notes several key markers:

In September, 2009, the *HealthLeaders* magazine cover story, "The Patient of the Future," focused on patient engagement. That December, the magazine named Dr. Sands and Dave to their annual list, "Twenty People Who Make Health Care Better," ranking just after Atul Gawande, MD, and inventor Dean Kamen.

In April, 2010, Lucien Engelen organized a TEDx conference in Maastricht, the Netherlands, and asked e-Patient Dave to speak. Lucien announced that because the conference was about health, a patient would be the first speaker. E-Patient Dave's talk, which ended with a chant of "Let patients help!" went viral on the TED talks website and rose to the top half of most-viewed TEDTalks. By now, his talk has received nearly 400,000 views, has subtitles in 26 languages, and has maintained its position in the top half of the most-viewed talks of all time.

In December, 2010, at the annual International Health Informatics Symposium, he and Dr. Sands presented the first talk after the opening keynote. It was a direct result of their having presented Medical Grand Rounds earlier that year at Beth Israel. Dave described the amazing experience of looking up into that Grand Rounds audience and realizing, with gratitude, that he was lecturing to the clinicians who had saved his life.

In 2011, the movement went overseas, with speeches in Denmark, Switzerland, Holland, Israel, and Spain, and a podcast for the *British Medical Journal* (in addition to TEDx Maastricht).

In October, 2012, the National Library of Medicine's History of Medicine Division reported that they had started capturing Dave's blog as a part of their permanent record.

The 2012 Institute of Medicine report, *Best Care at Lower Cost: The Path to Continuously Learning Health Care in America*, highlighted patient-physician partnerships with engaged and empowered patients as one of the keys to transformation of the U.S. healthcare system. Dave sees this as the ultimate validation from the medical establishment that e-patients and participatory medicine are not only real, they are genuinely important.

e-Patient Dave notes that as healthcare professionals first began to be aware of his story, they thought that either it was not real or e-Patient Dave was the first and only patient who was activated in that way. Today, he says, that's foolish because he was mainly talking about what other patients had been doing for years. As it became clear that he was not the only patient who was engaged and empowered, the tendency was still to rationalize it away—"It's only college-educated patients," or "It's only data geeks," or then "It's only Americans," etc. Gradually, however, it has become clear that the drive—and the ability—for patients to partner with providers around their own health is something that is universal and cross-cultural, enabled by a combination of forces that include the Internet, social media, and direct access to healthcare data. This realization also spawned the formation of the Society for Participatory Medicine, a 501(c)3 organization promoting the concept of participatory medicine, and on which both Dave and his physician, Dr. Sands, serve as directors and members of the Founders Circle.

Dave stresses that this is not just an issue for patients and patient rights. Providers cannot practice at the top of their license without access to complete data—sometimes data that is held by patients. He suggests that for real innovation to occur, it is important for more and better clinical data to be available to patients, and more and better de-identified data for researchers, and innovators. Dave believes that the next evolution is dependent on transforming the vast amounts of data we are creating into usable and meaningful information. The acronym he cites from computer science is DIKW—Data, Information, Knowledge and Wisdom. This means that data must be placed into context and that innovation is required to help review patterns and connections in the data, as well as to help people visualize complex systems.

When Dave looks back on this improbable story, it feels as if all of the elements of his history and skills have come together to enable him to do this work at this time. As the philosopher Kierkegaard once said, "We live our lives forward and understand them backward." As a child of the '60s, Dave always dreamed of changing the world. When we look at the changes in medicine that e-Patient Dave embodies and has popularized, we agree that he has accomplished that goal.

Case Study 8

From God to Guide

By Jan Oldenburg, FHIMSS (From an Interview with Lucien Engelen)

Lucien Engelen has dedicated his career to exploring what can happen in medicine when doctors move from "God" to "Guide," bringing their patients into the center of the circle, listening to them, and exploring what discoveries and opportunities arise from that intersection. He is the founder of REshape and the Innovation Center, programs of Radboud University Medical Center Nijmegen, dedicated to creating the decade of the self-empowered patient. He was the first Dutch Health 2.0 Ambassador.

In his role as the Chief Patient-Innovation Officer at Radboud Medical Center, Mr. Engelen uses patient-centered design principles to listen to patients and understand what they really need to help them cope with their diseases rather than starting with what the medical team thinks they need. In the process, he has uncovered something that seems as if it should be obvious, but is not—that stickiness and adoption are much higher when you build capabilities that people actually want. He notes that patients are the most underused resource in healthcare, and when you bring them into partnership as a part of the care team, radically different solutions are possible. Mr. Engelen and his team embody the principle outlined in Chapter 11, "always be innovating." Some of the innovations they have been a part of are highlighted next.

Crowdsourcing to Save Lives
Finding an automated external defibrillator (AED) when someone is suffering from sudden cardiac arrest can make the difference between life and death—but there is no simple, easy-to-use way to find them, just because nobody ever mapped them in a non-commercial, independent way nation-or-worldwide. Mr. Engelen started a program to crowd-source AED locations. The program (www.aed4.us) was so successful they were able to map the location of more than 17,000 AEDs and validate those locations through the Dutch Red Cross. They then developed a simple, free downloadable app that allows people to readily find an AED when it is critical. They are now trying to take the program international.

Peer Support for Young Adult Cancer Patients
One of the groups Mr. Engelen and his team talked to about their needs were young adult cancer patients between the ages of 18 and 35 years: AYA (Adolescent and Young Adults). They have specific issues that are different both from pediatric and

adult patients. When given the opportunity, the patients first asked the team to create a place at the hospital where they could hang out together and then asked if they could also build an online community where they could hang out virtually. One section of the online site includes hospital-provided educational content patients can access to check the accuracy of something said in the community, but it is first and foremost a community by the patients and for the patients. One of the surprising things to come out of the experience is that patients who are engaged in the community come into medical offices less often. Patients may make a telephone or face-to-face appointment with their physician, but they are more in control of their appointments and make somewhat different choices about what they do with those appointments. Mr. Engelen and his team are now trying to measure the magnitude of the effect and are in the midst of expanding it to other academic hospitals in the Netherlands.

Nothing about Me without Me

One of the things that illustrates Radbaud's "Patients as Partners" values is the way that they have handled patients' clinical data. They have just opened up all of their records, so that patients can see their full clinical records, including physician notes. Plans are that early next year, patients will also be able to download their records and choose with whom they want to share that data. "We at Radbaud University Medical Center only take care of patients in about 2 percent of their lives—the rest of the time they are on their own or with other providers." Providing them with full access to their records—including physician notes—acknowledges the patients' right to know what is happening with their bodies, with their care. Mr. Engelen's team has also experienced examples of the patient safety effect, in which patients who see their record are able to correct clinical errors, as discussed in Chapter 4.

As a part of this Innovation program, the team has just released an online video conferencing platform that is like Skype and completely secure and compliant for healthcare, available at: http:/en.facetalk.nl. Nurses and doctors at the hospital can—and do—conduct virtual conference calls about a patient's health and add the patient and the patient's primary care physician (PCP) to the discussion, without adding any hardware. It is one of many ways that Radbaud seeks to bring patients into the center of their care experience.

Social Health

Mr. Engelen is not shy about using his own experiences to understand how behavior change and social media affect healthcare. He told us (and has blogged about) the electronic scale that sends him his weight every day—and he started tweeting his weight each morning. The results were astonishing—people started sending him advice, "It looks as if you should have a light lunch today" and encouraging him when his weight was lower. He found that the peer pressure helped him. He also recently had his genome tested by 23andme and blogged about what he found. Chapter 9 discusses these and similar tools in greater detail.

The Future

Where will this lead us over the next 5 to 10 years? Mr. Engelen (who is also Faculty at Singularity University's FutureMed program) believes that more healthcare professionals will engage as patient guides, having real conversations with patients, rather than as the god-like figures they have been in the past. The results, he believes, will be more satisfactory for both patients and providers. In a study where a group of healthcare professionals and patients were each provided a description of two courses of treatment and asked to vote on which suited them best, 95% of the doctors chose the high dose, high cost, and high frequency of negative side effects option; 95% of the patients chose the option with low frequency, low cost and better quality of life.

Once you ask patients what is best for them, the conversation—and the outcomes—can change radically. In REshape Mr. Engelen appointed a Chief Listening Officer (CLO) to his team. She has no professional treatment relationship to the patient. Instead, she is there to listen: she talks to patients, caregivers, families, visitors, and staff about how Radboud can change their services for the better. That, along with patients' reviews, is input for all of the projects and their design teams. It serves as fuel to ignite change and innovation with patients as partners.

Case Study 9

Restoring Trust with Information

By Karen Colorafi, RN, MBA, BScN, CPEHR, CPHIT, and
Jan Oldenburg, FHIMSS (From an Interview with Richard Fitton, MB,
BS, MRCS LRCP, DCH, DRCOG MRCGP)

A famous story in the UK points to the lack of knowledge most patients have about what exactly is documented in their medical record. Harold Shipman, MD, might have killed as many as 260 of his patients with lethal doses of prescription painkillers before his arrest in 1998. He might have been caught sooner, but medical examiners, though alarmed at the number of deaths in his practice, were unable to locate any discrepancies in the patients' medical records. Only much later, a more thorough investigation revealed that Shipman had altered the medical records of his dead patients so the records would appear to corroborate their actual cause of death.

Richard Fitton is a general practitioner whose clinic is in Greater Manchester, near where Shipman practiced. He signed the cremation certificate for one of Shipman's patients, was interviewed by the police and testified at his trial. Dr. Fitton was a long-time advocate for providing patients with access to their medical records. He had started to advocate for patient access to their medical records as early as 1993 but faced an uphill battle. In the early years, local administrators and medical leaders reprimanded him for his advocacy and suggested that the very idea was mad.

Horrified by what happened with Dr. Shipman, and convinced that it would not have taken so long to discover if patients—and their families—had access to their medical records, Dr. Fitton began to provide his patients with full access to their medical history electronically in 2001; they were one of the first practices in the UK to do so. He is convinced that without Shipman's crimes, it would have taken far longer for UK patients to get access to their records online. Dr. Fitton said, "Shipman went unnoticed for so long by continually falsifying patient's medical histories. If patients had access to their records the way they have now at Hadfield, it is much less likely he would have got away with it."

When the National Health Service in England began putting patient records into a shared electronic database, Dr. Fitton wondered if he could do something to ensure the accuracy of his records before they were loaded into the national database. Fifty of his patients agreed to take part in the effort, and 31 returned comments. Most of those patients found nothing wrong with the records, and ten found one or more errors (30% error rate). How serious were those errors? It turns out that most of them were minor: an infant's birth date that was off by a year and one person's test results filed in another's chart. Some patients were troubled by things in the notes they did not want shared with the rest of the country, like dietary advice, history of mental disorders, terminated pregnancies, and a 17-year-old's request for contraception. Some patients disagreed with diagnoses like asthma or acne. Based on this experience, Fitton argued that enabling patients to see—and offer corrections to—their records is an important safety check: "Remember, the patient is the only person who attends every medical intervention that involves him or her, so the patient is the only person who will likely be able to spot reports and data in their record that are about another patient or are incorrect."

Ten years later, about 500 of Dr. Fitton's patients use the system. They report loving the access it provides. One patient wrote, "The system is great. It saves me a great deal of time and hassle. I can see all available appointments myself, rather than having a long telephone discussion with the receptionist. I can book appointments out of hours. If I wake up and know I need to see the doctor, I can make an appointment by 6:15 am. I can check that the hospital has sent reports to my PCP (primary care physician). I can download reports for other purposes such as an upcoming hospital visit or insurance purposes." Other patients echoed this sentiment, saying, "I can check to see if a letter has been sent, or see results online without having to ring the practice," and "This helped me accurately fill out travel insurance documentation and school immunization forms."

Patients also reflect that the system is useful for more than the convenient self-service features. As one patient said, "I am a firm believer that to a large degree, one is responsible for one's own well being and therefore one has to have some degree of self-knowledge in the medical sense. Therefore, my intention and use of the Medical Records Access process was, and is, not so much to make or vary appointments as to better enable me to understand the cause, make-up, and progress of the chronic diseases that I suffer from, and their various medicinal treatments, so that I may be better able to understand and discuss these with my GP [general practitioner] at my half-yearly blood/urine tests and review appointment. As an example, by extracting and plotting my blood levels, I am able to see that although there is spot variance over the last few years within the band, the trend is not worsening. This allows me to have an intelligent discussion with my GP to better understand his view and recommendations."

Dr. Fitton is now working on global health education and continuing to advocate for the rights of all patients to have access to their own data.

Case Study 10

The E Is for Engagement*

By Susannah Fox

What if we redefined the Quantified Self movement to include everyone who keeps a pair of "skinny jeans" in their closet? What if the 85% of U.S. adults who own a cell phone understood that it is potentially a tool for health tracking? What if everyone designing healthcare tools first talked with patients and caregivers about what they need, instead of making assumptions, without input?

These were the themes of two talks I delivered recently, first at Stanford Medicine X and second at the Connected Health Symposium. Following is what I said:

The "e" in e-patient stands for engagement.

First, an acknowledgment. Tom Ferguson was an MD who believed he saved more lives by not practicing medicine in the traditional sense, but by forging a path we have come to call participatory medicine. He was my mentor for the six years I knew him before he died in 2006, and he continues to teach me as I review what he shared with me.

Tom coined the term *e-patient,* which many of us use today to describe people who are engaged in their health. He saw clinical care as one piece of life's puzzle. A person's own motivations, experiences, and values form other pieces of the puzzle. Tom recognized that clinicians should not try to fit the pieces together for their patients, but rather act as a guide when life intersects with medicine.

Think, for a moment, about the puzzle pieces of your own life. Think about the pieces you know well and the ones whose edges are blurry or indistinct. How can you go about learning the shape and contours of those mysterious pieces? And which ones intersect with health and medicine?

Now think about how a clinician might view your life's puzzle. How can you help them to see how the pieces fit together? Which pieces do you want to keep private and even hide from view, because they are too personal or scary or embarrassing to reveal? What self-knowledge do you wish you had and do you want to share it with anyone other than yourself?

These are the sorts of questions that the Pew Internet Project and California HealthCare Foundation considered as we went into the field with our latest health

---

*Reprinted with permission from the Pew Research Center's Internet & American Life Project.

survey. I'll give you a sneak preview of the results related to engagement, because I think our field is moving too fast for me to wait for publication.

We decided to focus a section of our survey on self-tracking, which is really self-discovery, a basic component of engagement.

Looking at my own life's puzzle, I don't think of myself as a self-tracker. I don't own a FitBit or a Zeo. I run, but I don't track my mileage or my pace. I don't even own a scale, relying on that time-honored measurement—do I fit into my skinny jeans? Raise your hand if you own a pair of "skinny jeans." But I realized that, in this case, I represent what we Internet people call "the normals." The vast majority. The opposite of the early adopters. The people you may want to reach. So I set out to take some measurements.

We found that 60% of American adults track their weight, diet, or exercise routine. One third of American adults track health indicators or symptoms, like blood pressure, blood sugar, headaches, or sleep patterns. One third of caregivers—people caring for a loved one, usually an adult family member—say they track a health indicator for their loved one.

Putting that all together, 7 in 10 American adults are self-trackers. But guess what? Half of them are tracking "in their heads." These are my people. I'm calling this group the "skinny jeans trackers." In addition:

- One third of self-trackers use a notebook or journal.
- One fifth of self-trackers use an app, a device, a spreadsheet, or a website.
- Half track on a regular basis.
- The other half track when something changes, when something comes up and triggers the need to track.

My question for you is: How are you going to create something that is as easy to use, as simple to fit into someone's life as keeping track in their head? That's your goal. Those are the people you may want to reach. They are self-tracking "sleeper cells" that you want to figure out how to activate.

In our health surveys, we ask people if they are living with diabetes, lung conditions, heart conditions, high blood pressure, cancer, or the catch-all "any other chronic condition."

I took a look at the group of adults who are most likely to need to track a health indicator or symptom—those who are living with two or more chronic conditions. Sixty-two percent of adults living with two or more chronic conditions is self-tracking.

This is the kind of data that you can use to map the frontier of health and healthcare. That's why I call myself an "Internet geologist." You need to know the history of the land you live on. You need to understand the patterns of current technology use, so you can build something that will last. Don't build your house on sand, build it on bedrock. Lay down paths where people already walk.

So, where are the paths online? How can you follow people—or redirect them to a healthier lifestyle? The path that most Americans are traveling is on their mobile devices. Eighty-five percent of U.S. adults own a cell phone (or as I like to call it, my tracker).

Paths specific to health:

- 19% of smartphone owners have a health app on their phone.

- Most popular categories of apps include exercise/fitness, diet/food, and weight.

- Weight Watchers is an example of a successful program that includes and encourages self-tracking. They've had 10 million downloads of their apps, which are integrated with their website and in-person support group program. The secret to their success is the integration, which enables engagement.

Think about it:

- We have a big group of people who are living with obesity, high blood pressure, diabetes, and other conditions, which might improve if they engaged in self-tracking.

- We have a big group of people who are doing some self-tracking—and there's even some overlap in those groups. But it's mostly in their heads or on paper.

- We have a big group of people who have cell phones, with the potential for an upgrade in how they self-track.

- But right now it's just a small group who are using apps, websites, devices, or spreadsheets to do any self-tracking.

Given the rise of smartphones and online tools, should we expect to see the percentage of self-trackers go up? Should it be our goal to see 100% participation in self-tracking among people living with chronic conditions? What about among caregivers? And the general public?

Note that last bullet: look for outliers. Watch the mainstream, but also keep an eye out for pioneers.

For example, there are some intriguing small groups, like the men who have a menstrual cycle app on their phone. They might be tracking fertility, either trying to make a baby or avoid it. I am sure we can all think of other reasons to track menstrual cycles.

Speaking of which—this is for the ladies—how many of you have showed up for your annual exam and not been ready with the date of the first day of your last period? Your clinician expects you to self-track and is exasperated if you don't keep good records and share it with them. Or at least mine is.

People living with diabetes face the same challenges around the expectation of self-tracking. As a blogger who goes by @txtngmypancreas wrote recently: "As patients, it's not enough that we have to live with the disease itself. We have to live with the data management as well." Who is providing the tools to help them? And what is the expectation about their sharing that data?

Which brings us to the topic of sharing in general. Going back to our general self-tracking survey questions, we asked people: Do you share your self-tracking records or notes with anyone, online or offline? Two thirds say no, they do not share. One third say yes, they share. Of those, half say they share their self-tracking notes with a clinician. The other half share with a family member, friend, or member of a group.

What does that tell you about clinical engagement with self-tracking? Are doctors and nurses aware of how many people are self-tracking? Are they maintaining a don't ask, don't tell policy? Who are the patients and caregivers who are sharing, who are showing and telling? How are they different from those who do not share?

Katie McCurdy is an e-patient who created a medical timeline so that she could discover things about her own health and show the puzzle pieces of her life to her clinicians. She is one who decided to share, but remember and respect the two thirds who do not share their notes with anyone.

Think back to where I started—which puzzle pieces of your own life do you display publicly or talk about with clinicians? Which pieces of your life's puzzle do you keep to yourself? How should that be reflected in our discussions today about self-tracking?

One of the first lessons I learned from Tom Ferguson was to never assume that I knew the full story of someone's life. It was Tom who encouraged me to do fieldwork in patient communities, listening and interviewing people about how they use the Internet as they pursue health. No matter what your connection is to healthcare, you can benefit from listening to patients and caregivers.

Tom drove this lesson home by recommending a book by the anthropologist, Diana Forsythe, entitled *Studying Those Who Study Us*. In one chapter she describes how, in the 1990s, she did some fieldwork in an artificial intelligence lab that was asked to create an information kiosk for newly diagnosed migraine patients. The idea was that patients could walk up to the kiosk, punch in questions, and get some answers before or after they saw their doctor. It was a nice idea, ahead of its time in some ways. But when it launched, it was a failure. Patients were polite, as they often are, but it wasn't useful. Why? Because the kiosk's designers had not asked patients what they wanted to know. They relied on an interview of a single doctor to tell them what he thought patients should want to know. As Forsythe wrote, "The research team simply assumed that what patients wanted to know about migraine was what neurologists want to explain."

The mismatch was complete. The kiosk failed to answer the number one question among people newly diagnosed with migraine: Am I going to die from this pain? It's an irrelevant, even silly question from the viewpoint of a neurologist, but it is a secret fear that people may have felt comfortable expressing to a kiosk, but not to their clinician. A secret piece of their puzzle.

How many of us have asked Dr. Google a secret question? I know I have. The open search box is so inviting, especially on our mobile phones, when the intimacy is multiplied.

Think about how you can use that in your work. Think about how you can avoid the assumptions that will torpedo your project. As Forsythe taught us: There is so much potential for greatness, so much potential for impact on people's lives, if we can just "design for what could be."

The Pew Internet/California HealthCare Foundation survey asked people about the impact of self-tracking on their lives. I'm going to close with these numbers, because I think they represent "what is." I want you to go build "what could be."

• 34% of self-trackers say their data collection affected a health decision.

- 40% of self-trackers say it has led them to ask a doctor new questions or seek a second opinion.
- 46% of self-trackers say it has changed their overall approach to health.

Let's help people solve their own puzzles. Let's create tools that engage people and respect what they want out of life. Let's map the frontier of what could be.

Case Study 11

The Power of Information and the Partnership of Trust

By Amir S. Hannan, MBChB, MRCGP, and Glen Griffiths, MSc, MBCS

We are a small practice in the UK, with just under 12,000 patients registered with us. We treat patients as part of a well-developed, primary (community-based) care service. We implemented an electronic health record in 2000, storing all patient consultations conducted within the practice, as well as home visits, lab test results, and letters from other healthcare providers. In 2006, we enabled patients to access their full electronic health record via the Internet. Our practice covers the area where Dr. Shipman, mentioned in Case Study 9, practiced and used falsification of medical records in an attempt to conceal his crimes.

Today, overall, 16% of patients (1,917 in total) have registered for online access. These include 22% of patients with diabetes, 26% with anxiety and/or depression, 27% with cancer, and 44% of patients taking methotrexate (requiring regular blood test monitoring). We have patients from birth to over 90 years of age who have signed up, including patients with learning difficulties, caregivers and even those for whom English is not a first language.

We Have Not Had a Single Problem

Our system, www.htmc.co.uk, was developed in-house with patients and clinicians. Its purpose is to help patients understand what records access means for them and to provide a safe environment to find trusted health information, as well as locally-derived guidance. The site includes guidance provided by patients, too. As challenges arise, we produce information with links to other material to help to improve individual health literacy, which patients themselves describe as empowering. This is now recognized nationally as supporting the power of information and producing a partnership of trust between patient, clinician and organization. (See http://www.htmc.co.uk/pages/pv.asp?p=htmc0395.)

How Do Patients Get Access to Their Full Electronic Health Record?

Patients are encouraged to sign up for the service through a wide variety of communication channels. Information is provided to help them understand the issues surrounding access to individual health records, including demonstrating the capability with a test patient record prior to their signing up for the service. We explain the need for privacy, confidentiality, security, and how to share information responsibly. In order to gain access, patients first have to take a test to show they are capable of handling personal information and its consequences. The practice also checks to ensure there are no reasons why the patient should not have access (e.g., in the case of severe mental health problems). The explicit consent process is continual and can be revoked if patient access is deemed unsafe or if a patient chooses to stop access (see http://www.htmc.co.uk/pages/pv.asp?p=htmc0378).

What Benefits Have Patients Identified?

Patients describe improved relationships with care providers, the ability to better manage and track their own, and in some cases, their family's health matters, a more thorough understanding of their own health, the ability to identify errors and omissions—in some cases preventing unnecessary tests, thus saving time and money. Patients also describe being able to make plans around the time of death with compassion and dignity.

What Does the Future Hold?

We are still developing ideas, continually innovating and aiming to enable all our patients to benefit from the service. We are encouraged by Kaiser Permanente's approach, as well as others as we break down barriers and empower patients, clinicians, and organizations to deliver better, safer, more cost-effective care— closer to home—with an improved experience all round.

Case Study 12

Using *iTriage*® to Engage Healthcare Consumers

By Peter Hudson, MD

When patients are actively involved in their own healthcare decisions, they are more likely to work collaboratively with their healthcare providers. This active role is vital for those suffering from acute conditions who have immediate needs requiring navigation of the healthcare delivery system. It may be even more important for those living with chronic conditions. Empowering patients with reliable and readily-available heathcare information is a crucial step toward achieving increased patient engagement.

After seeing tens of thousands of patients struggle with access to the right healthcare information at the right time, my partner and I, both trained emergency physicians, developed a free mobile healthcare application called iTriage. In our practice we recognized there was a groundswell of consumer need for quick, actionable information that would solve a critical problem: getting the right information at the right time to enable patients to make better, and more informed decisions about what kind of medical help they need. Solving this need allowed us to go from treating 50,000 patients in 20 years as medical professionals to providing access to treatment for seven million patients in three years as software developers. That's the power and scalability of today's technology.

Mobile Solution Brings Connectivity

Years ago, patients trusted their doctors explicitly, waited for appointments patiently, and took medications without question. With today's information explosion and 24/7 mobile access, patients now demand both reliable health information and a direct connection to their providers. iTriage fills this gap between medical professionals and the patients who need their help. By giving context to medical information and then quickly leading consumers to the closest and most appropriate provider, iTriage facilitates consumer and patient engagement in many ways.

First, the symptom, disease, and medical information that are the core of iTriage empower consumers with a taxonomy of information for quick treatment decisions. This information can also provide people with a foundation of understanding that medical providers can build on, thus increasing doctor/patient engagement within the time constraints of a medical practice.

Second, valuing a patient's time has been at the forefront of medical provider news over the past several years. iTriage began addressing that patient need when it first partnered with a hospital system in 2009 to roll out emergency room wait times for mobile devices. iTriage has taken this a step further with mobile and online appointment setting for patients and their medical providers. People appreciate the convenience of making appointments anywhere and at any time, while medical providers appreciate a new channel to engage with patients.

Third, an iTriage feature that allows consumers to notify healthcare providers of medical emergencies and their symptoms breaks new ground in patient engagement. *ER Check-in*™ and *Urgent Care Check-in*™ provide a connection between incoming patients and their providers at a time when stress is at its highest and outcomes require rapid and efficient care. This additional level of communication improves healthcare delivery by giving providers a "heads up" for additional staffing needs and helps doctors be better prepared for the type of emergency that will soon arrive.

Results from an Effectively Deployed iTriage® Solution

A hospital system known for offering top quality care identified a growing need for differentiation in their local marketplace. To increase patient engagement, the hospital system wanted to go beyond providing the highest level of care for residents, newcomers, and travelers by heightening their top-of-mind community presence with a new level of patient connectivity.

Hospital administrators selected iTriage's consumer/patient engagement solution, providing ER Check-in and Appointment setting. Knowing that teamwork would be key in this roll out, a hospital team and iTriage developed a marketing plan to maximize exposure and increase adoption of the patient engagement solution. Just 34 days after signing the contract, the hospital system went live with *ER Check-in*™ and used the following tactics to drive awareness:

- Identified a hospital system iTriage champion to create internal ownership of the solution.
- Developed a competitive hospital facility program to influence engagement.
- Distributed a press release and executed a marketing campaign to create awareness within the community.
- Began personal, onsite communication at every facility.
- Implemented a physician liaison program to drive clinical referrals for ER Check-in.
- Embraced digital strategies to further adoption, including hospital website and mobile site marketing.

Hospital executives ensured positive momentum throughout the implementation of the iTriage solution by obtaining buy-in from the entire hospital team, including executives, physicians, and staff. This teamwork approach to the iTriage solution resulted in 215 ER Check-ins within the first 30 days, which gave the hospital system its first glimpse into how iTriage can enhance the patient experience with their facilities. As the system rolls out further, the solution will include appointment setting for physician partners. The early awareness created in the community with ER Check-in will transfer effectively to a broader engagement strategy.

Case Study 13

Meaningful Use Implementation and Patient Activation

By Jan Oldenburg, FHIMSS (From an Interview with Stasia Kahn, MD)

Dr. Stasia Kahn is passionate about achieving a fully operational healthcare system in her lifetime. She defines "fully operational" as having a practice that:

- Is adept at treating acute and chronic conditions, as well as fostering wellness.
- Acts on plans of care for individuals and is proactive in improving care for populations of patients.
- Ensures that relevant health information is available at the point of care.
- Implements technology wisely.
- Allows consumers to access their clinical data.
- Is interoperable with healthcare institutions in order to share protected healthcare information.
- Has a sustainable business model.
- Achieves harmony in work flow.

Dr. Kahn's desire to achieve a fully operational practice has led her to be both an early implementer for Meaningful Use Stage 1 and actively engaged in experimenting with, blogging about, and promoting the use of healthcare IT to improve practice and the health of patients. Her blog is www.emrsurvival, and it is full of practical tips about the advantages—and the pitfalls—of implementing an EHR and attesting to Meaningful Use. Her 90-day attestation was done in 2011 in her previous practice; she opened a new practice in April 2012 and is doing her 2012 attestation using data from two practices and two different EHRs. Her experience suggests that there are real benefits in clinical practice to implementing Meaningful Use, and that the incentive payments are a welcome offset to the cost of the implementation.

Dr. Kahn's first practice did not implement the patient portal because it was an add-on cost that they simply could not initially afford. It was a bonus that the EHR she selected for her second practice offers the patient portal free for the first two years. Dr. Kahn has been highly assertive about getting patients signed up for the portal and actively using it. In the first six months after launching the patient portal, Dr. Kahn has signed up 60% of her patients. Her whole staff is trained to get patients signed up and to encourage portal use. Dr. Kahn promotes the portal by explaining to each patient that they can view results from preventative exams and tests including pap smears, colonoscopies, mammograms, dexascans, screening lab tests, and immunizations. Additionally she points out that each one of their lab test results are viewable on the portal, so patients do not have to write down their results on paper when the nurse calls to discuss the results. The new practice has implemented additional functionality offered by the new software including setting up alerts for appointments, preventive services, and repeat chronic disease visits so that patients can receive emails and/or phone calls reminding them to come in. She offers medication refills through the portal, as well as eConsults. Her website posts instructions for patients on how to set up and complete an eConsult, though uptake on the feature has been lower than originally expected. Dr. Kahn attributes this to the fact that it is still difficult to get insurers to reimburse for virtual visits.

Dr. Kahn is already planning for Meaningful Use Stage 2. The component that makes her most uneasy is the requirement for data exchange with other physicians for referral and transitions of care. There is not yet a simple way for physicians on different electronic medical records to find one another and set up the appropriate permissions and passwords to exchange data with one another, even when health information exchanges (HIEs) are involved. It would be extraordinarily helpful, especially in small practices, to have a physician locator service that crossed regional HIE boundaries, coupled with a nation-wide secure email program that easily enabled secure messages to be exchanged between all the doctors involved in a patient's care. Right now, it requires physicians to network with each other to set up all the appropriate online relationships to enable sharing data in situations where the ability can save a patient's life.

Despite some of the difficulties of being an early Meaningful Use adopter, Dr. Kahn believes it has brought real and tangible benefits to her practice, helping her to achieve the fully operational practice described earlier.

Case Study 14

Moving the Point of Care into Patients' Daily Lives via Text Reminders

By Brad Tritle, CIPP, and Karen Colorafi, RN, MBA, BScN, CPEHR, CPHIT (From an Interview with Joseph C. Kvedar, MD)

Two of the hallmarks of mobile health are accessibility and scalability. This can be seen in the many ways that mobile technology comes alive in the healthcare arena, providing functionality on devices that are nearly ubiquitous and designed for ease of use by all populations. When used effectively, applications on mobile devices can improve patient engagement, adherence, and ultimately, outcomes. These are the concepts that have driven the Center for Connected Health to move healthcare out of the doctor's office, into the lives of patients and consumers over the last 18 years.

Most notably, the Center designed a program to connect with an underserved population using text messages. They targeted a specific population of patients who had complications during pregnancy and poor no-show rates at the doctor's office. Young mothers, ranging from 14- to 32-years-old, were sent regular text messages (one to four per week) reminding them of upcoming office appointments but also noting predictable milestones like the date of the next ultrasound or a reminder to start shopping for car seats. The outcome was that women who received the text messages received the recommended levels of pre-natal care 9% more than women who did not receive the messages.

Examples of the messages included the following:

"Your OB team wants to remind u that u can call us anytime @ (xxx) xxx-xxxx. Stay on the line and don't forget to tell us you're pregnant."

"Hi, it's your OB team. We want to make sure u have a plan to get to the Birthplace. Let us know if we can help."

They report that 100% of participants reported reading all or most of their messages.

Center Director Joseph Kvedar, MD, explains that "OB care is an interesting place to start a program like this, because once you know the due date, you can predict all kinds of behaviors and events that are going to happen." From a technical standpoint, the predictability makes it easy to build an algorithm that fires text messages off automatically.

It's also a great place to start from a business perspective. Newer models of care such as accountable care organizations (ACOs) and patient-centered medical home (PCMH) practices must hold on to their patients in order to make money because, for example, were a pregnant patient to see a competitor, the ACO has to pay the bill. Texting not only encourages loyalty during the pregnancy but also makes it more likely that a woman who is satisfied with the care she receives during pregnancy is likely to bring the whole family to your practice for years to come.

Having experienced tremendous success with texting prenatal patients, the center is working on similar programs that target patients with addictions and diabetes. Patients who struggle with addictions can be reminded via text message of upcoming blood and urine drug tests and can be sent encouraging, motivating notes about staying sober. Patients with diabetes complete a questionnaire that identifies their readiness to change. They are given a pedometer that feeds back a total step count, which enables the center to customize the messages sent to the patient encouraging them to reach a higher activity level.

Dr. Kvedar explains that one of the most intriguing things about text messaging is its intimate quality. People still open 90% of their text messages, and they do it immediately, unlike emails which have poor read rates. In a country which boasts more cell phones than humans (U.S.), the Center for Connected Health will continue to use text messaging as a strategy to engage consumers.

Dr. Kvedar is also fascinated with the potential of social media in healthcare. He said, "The fields of behavioral psychology and behavioral economics have a lot to teach us about reaching out to people, about what pushes someone's buttons. When we know what motivates you, what is really important to you, we can customize the messages that you receive for maximum results." He went on to explain that "Medicine has been focused on physiology—plaque in the heart, ulcers in the stomach, damaged cells in the liver—we view everything as a physiological problem to be solved with a physiological response. Now we are relearning that this process is subject to all kinds of things like emotions, engagement, readiness to change, and an interest in being healthy. The next phase of our work involves capturing more and more of this information and using it effectively to customize treatment. Imagine a day when we can intuit the kinds of messages that will be relevant to a particular patient based on patterns of behavior."

Learning what motivates a particular person appears to be the very cutting edge of science and the nexus of mobile technologies, patient engagement, and healthcare.

Case Study 15

New Models of Education for Pediatric Diabetes

By David W. Moen, MD

Knowing more answers than the person next to me is what got me into medical school and propelled my performance on exams and rotations as I progressed through my training. However, leadership in innovation requires that we admit what we don't know and engage others in the process of achieving goals that matter most to them.

Patients and families spend more time living with their conditions than we do. Patient engagement remains our biggest challenge if we are going to truly affect the health of the populations we serve. During the period when I was leaving the role of emergency department (ED) Medical Director and staff physician and transitioning to the role of Executive Medical Director of Innovation for the Fairview system in Minneapolis, Minnesota, we used a design and innovation framework to find the best way to equip families with a child newly diagnosed with diabetes to learn the skills necessary to live full lives. That work has evolved to a functional online pediatric diabetes learning community that supports the work of clinicians and researchers at the University of Minnesota Pediatric Endocrinology Department.

As a practicing emergency physician, I experienced the gaps in our healthcare delivery system during every shift. Three areas emerged as ripe opportunities for innovation to advance our efforts to more effectively and efficiently engage patients and families. Those areas included communication, learning, and behavior change. Emergency visits often occur because patients and families do not have access to the right person in the right place at the right time. Patients who understand their conditions and have confidence in their self-management skills have lower health system usage and better patient outcomes. Behavioral choices determine approximately 40% of our health, and poor choices by patients generate huge costs today and into the foreseeable future.

In October 2008, Brandon Nathan, MD, and his team at the University of Minnesota agreed to use a design and innovation process to help us better understand how to engage patients and families. We agreed that the Institute for Healthcare Improvement's Triple Aim of improving the patient experience, improving the health of populations, and reducing the per capital cost of healthcare would be our team's end game, and we aligned it with our patients' and family's desires.

We opened an online community to patients and families with diabetes and asked them to weigh in as we strove to improve their clinical outcomes, contain costs, and improve their experience of living with their disease (both when they are seeing us and when they aren't!).

The following are the most important themes that we heard from families in the online community:

1. Better access to the right person at the right time is critical and doesn't happen reliably today.
2. The Internet is full of information—so much that it is overwhelming and sometimes frightening!
3. We trust our doctors and nurses and the information provided by the clinics where we see them, and that trust is essential in addressing the fears we have as we cope with a chronic disease.
4. We often have many questions that we forget to ask when we see our doctors. Can we ask questions more often and can we share the answers with others?
5. There is a lot to learn and treatment recommendations change often. How do we best keep up-to-date with accurate information?
6. We (patients/families) learn a lot in the course of adapting to an illness in our family. Can we share that with others?
7. Can we (patients/families) have access to and contribute to research? We want to help progress toward a cure!

After a three-month period of online dialogue, we supplemented our learnings by hosting face-to-face focus groups. It was very clear that people had access to technology and were willing to help us design better tools to increase the patient/family/clinician connection. All of us saw an opportunity.

The team decided to change the learning model first. Patients, families, and educators reported that repeated lectures of didactic content were boring (for everyone!). It was clear that information was repeated over and over through the course of a day. Patients and families desired accurate information endorsed by providers. They were also willing to contribute what they had learned—so we decided to host a contest. We invited families with a diabetic child to come to the Student Union at the University of Minnesota to share their stories on video. We offered a $25 gift card as an incentive. We also offered a $50 gift card to the patient and family that produced the most informative and creative video on their own. To our delight, 29 patients and family members came to contribute their stories and a nine year old and his father produced a winning video about *living* with diabetes.

The clinical team determined the learning benchmarks needed to successfully manage diabetes. The video session was purposefully designed to encourage patients and families to address those themes. We learned that parents of children with diabetes were very frightened early on, and we realized that people who are frightened don't learn. Most of these families had relatives or friends with diabetes growing up. Their vivid memories included people who experienced blindness, kidney failure, amputations, and early death. Those images fueled fear that impeded learning and engagement.

The winning video showed a nine year old playing soccer with his friends, lounging on the beach, and playing piano—all while living with diabetes. He demonstrated with confidence the safest way to do his insulin injection. His video proved popular with patients and families and we also showed it to a few skeptics along the way (you may run into a few of those as you innovate!).

To test the online community offerings, Dr. Nathan's team measured nine newly-diagnosed patients' and families' achievements of learning objectives. Educators didn't give lectures; instead, their role evolved to that of engagement managers. Patients and families used the online community as their learning resource. In a small study presented as a poster at an international juvenile diabetes conference, Dr. Nathan shared results that demonstrated that all families achieved their learning benchmarks. Families also shared that this experience positively influenced their perceptions of their clinical team, enough to influence their decision to choose this practice as their specialty provider.

Providing access to this resource improved learning and decreased the number of phone calls that interrupted nurse educators during the day. Though producing proactive communication tools takes time and improving access intuitively feels as if it should require more work, our team's experience reinforces what we have learned in other venues: access to the right person at the right time actually saves time and improves practice efficiency. Barriers to access cost all of us money and time.

The next phase of innovation, underway now, is designed to explore ways to improve engagement in the least-engaged and highest-risk patients. This population consumes tremendous resources and suffers significant morbidity over time. Dr. Nathan's team is using virtual care delivery tools and web-enabled glucose monitoring to manage these high-risk patients more proactively. Rather than waiting for quarterly visits, the team is touching base more often. This study is designed to measure the impact of this proactive, virtual management model as a more effective way to supplement face-to-face visits. The payment barriers are broken only when we push forward to demonstrate that more frequent, relevant contact actually lowers cost.

Important transformational themes have emerged in the work of our pediatric diabetes team with their patients and families. These themes are supported in other innovative work occurring around the U.S.:

- Relevant responsiveness fuels all aspects of the Triple Aim.

- Trust is built over time through collaboration with patients, families, and care teams. Trust and knowledge lead to actions necessary to change behavior and cope with chronic disease.

- Today's predictive models are helpful, yet insufficient. Understanding the geography of a patient's life is important. Care models that meet patients in their geography have much higher chances of success.

Good luck and keep innovating. The best solutions lie in the hearts and minds of patients and clinicians!

Case Study 16

Online Therapy and Patient Engagement

By Jan Oldenburg, FHIMSS (From an Interview with DeeAnna Merz Nagel, LPC, DCC, BCC)

Before DeeAnna Nagel began practicing online therapy, and before she co-founded the Online Therapy Institute, she was a patient, and an e-patient. In early 2000, she was nearly incapacitated with a mysterious lung disease. Her medical journey took her from specialist to specialist, each baffled about her symptoms, while she got sicker and sicker. She discovered an online forum for lung disorders, where she found a disease that matched her symptoms and information about the tests that would prove it. She went to her doctor and explained what tests she wanted and what disease they were testing for. Her doctor was skeptical and patronizing, but did the tests, which confirmed the diagnosis of sarcoidosis Ms. Nagel had found through the online forum.

During the next year, Ms.Nagel was unable to work outside of her home while she recovered from her disease. She began working as an online therapist with eClinics, Help Horizons, and Here to Listen, all arranged so that a therapist can make him/herself immediately available to a client for drop-in chats. With the benefit of hindsight, Ms. Nagel noted that these are not recommended conditions for an e-therapy relationship, as it means you are doing triage with a patient before you are established as a trusted and contractual clinical relationship.

As a result of this experience, however, Ms. Nagel became intrigued by the opportunities for online therapy, but also was deeply aware of the need for an ethical framework that would govern its practice and the use of online tools within a therapeutic framework. Peers from the International Society for Mental Health Online came out with a set of original guidelines in 2000, but Ms. Nagel and colleagues felt there needed to be more. The American Counseling Association (ACA) issued their last code of ethics in 2005. While the code covered most issues related to online therapy, there have been significant changes in cyberspace with the advent of Web 2.0. For example, social media was not fully in existence in 2005 and, therefore, the ACA code did not address how social media can impact a therapist's work and professional presence. Their next code of ethics will not be issued until 2014. To fill the gap, Ms. Nagel and a colleague started the Online Therapy Institute. They created a set of guidelines and issued them as a starting point for others practicing in the field. Ms. Nagel is clear that the standards of practice are still evolving and will continue to change as practices and technologies to support them continue to evolve.

Many therapists who practice in traditional face-to-face settings are finding that communication that formerly happened by phone or in the context of a session has moved to email. As a result, they need to determine how to weave online tools into the fabric of face-to-face practice. The American Psychology Association's (APA) Center for Workforce Studies conducted a study that showed that overall email use with clients for service delivery more than tripled among practicing psychologists from 2000 to 2008, with approximately 10% of those sampled using it weekly or more in 2008. Practitioners' use of videoconferencing with clients, while still rare, increased from 2% to 10% among survey respondents during that same time period.[1] This brings a host of both opportunities and problems. Even therapists who do not think of themselves as practicing "online therapy" need to determine such things as when an email exchange becomes an e-visit and requires payment, when an issue can be addressed by email and when it needs to come back into a session, what their response time to emails will be, and how they set boundaries and expectations when email is a part of the therapy mix. An online code of ethics needs to address these types of issues both for therapists who use online tools as an adjunct to their practice, and for those who primarily practice using online tools.

What does online therapy look like? It can take many forms and both therapists and patients gravitate toward the mix that fits them best. Ms. Nagel loves written language and believes that writing can be a powerful healing experience for patients, so she gravitates toward email and chat. She notes that for her patients the experience is rather like keeping a journal, but with a listening ear on the other side. Her Employee Assistance Program (EAP) work is all email-based. Sometimes her face-to-face clients will see her online, so online therapy becomes an adjunct to in-person consults. She uses video conferencing, chat, email, and phone in her work. For her, personal video is the least-preferred option, in part because the still-jerky head movements and lack of direct eye contact make it difficult to "read" her clients.

One of the important lessons that Ms. Nagel has learned is that whenever technology is a part of the mix, there will be glitches. The problems may be as diverse as emails that are lost in cyberspace, chat or video Internet connections that don't work when they need to, messages that are garbled because of fat fingers on the keyboard or inappropriate auto-corrections, simple misunderstandings based on the written word, or timing issues based on asynchronous communications. Therapists need to prepare themselves and their patients for these kinds of problems and create back-up plans to address them. Technical problems cannot entirely be avoided, but a clear communication plan and agreed-upon strategies for dealing with problems will go a long way toward mitigating their effects.

Ms. Nagel believes that several things are required for e-therapy to become more mainstream. One is a reimbursement model that consistently acknowledges and enables payment for therapy using online tools. She suggests that additional studies comparing the effectiveness of online therapies to in-person therapies will be required to create the impetus for a consistent reimbursement model and consistent way of coding for online therapy. Several studies and meta-analyses have shown that the outcomes are the same or better than in-person approaches and that it can be delivered at lower costs. Her fear, however, is that as it becomes mainstream, organizations will expect therapists to use online tools without any training or guidance. Ms. Nagel is advocating for a continuing education requirement that involves taking a course that will provide guidance about online therapy and teaches therapists an ethical framework for online engagement.

The other dimension that Ms. Nagel suggests will be required for online therapy to become mainstream is more and better tools. Advances in secure person-to-person video will certainly enhance the options available to therapists and patients alike. She and her colleagues dream of online therapy tools that will create a multi-dimensional experience, in which patients could securely send and receive messages, watch a video, post pictures or status updates, and perhaps interact with other patients.

Ms. Nagel also notes that there is tremendous research and energy going into exploring Avatar therapy and the use of virtual worlds for therapy. She notes that the Online Therapy Institute does a monthly meeting in Second Life, and that InWorld Solutions has a Health Insurance Portability and Accountability Act (HIPAA)-compliant platform for virtual world therapy. However, as she notes, many therapists feel like "digital immigrants" struggling to make sense of their options and how to interact in a virtual world. The capabilities are easier for 5 year olds to use than 55 year olds, and full adoption may need to wait until the next generation comes of age.

The potential is huge for online therapy to become an effective tool for delivering mental health support to a broad audience in a cost-effective way, and it is well-worth an investment in technology, training, and ethics to achieve the potential.

REFERENCES

1. Novotney A. A New Emphasis on Telehealth: How can psychologists stay ahead of the curve – and keep patients safe? *Monitor on Psychology*. June 2011; Vol. 42, No. 6. http://www.apa.org/monitor/2011/06/telehealth.aspx. Accessed November 6, 2012.

Case Study 17

NeuCare Family Medicine, a Direct Primary Care Practice

By Jan Oldenburg, FHIMSS (From an Interview with W. Ryan Neuhofel, DO, MPH)

Dr. Ryan Neuhofel, or Dr. Neu as he likes to be called, is a family practice doctor in Lawrence, Kansas. He practices a new kind of medicine—Direct Primary Care. It is much like concierge medicine but at a fraction of the price. Dr. Neu's patients pay him a small stipend monthly in exchange for services such as quick phone calls, lab and diagnostic tests, online personal health record, pharmacist counseling and group education classes. His patient population is much closer to that of a safety net clinic than a concierge practice: 60% to 70% are uninsured, and the average household income is $30,000. The prices for services are all posted up front. He practices high-touch medicine, offering house calls and finding patient discounts on related services, but he also practices high-tech medicine, using a free practice management system, offering his patients access to their information via a portal, and responding directly via emails and text. He has also built his own website and created a number of his own patient videos.

Even during medical school and residency, Dr. Neu was frustrated by traditional practice management approaches and the layers of overhead he experienced working with billing and coding events. He decided early in his residency that he wasn't going to have an insurance-based practice and carried through when he opened his practice in November 2011. His goal continues to be to make his practice as lean and efficient as possible, while providing a higher level of care and service than in a traditional practice. Dr. Neu started his practice straight out of residency with no loans or credit. He saved the money he made moonlighting in emergency rooms (ERs) and used it as his only source of start-up capital. He says, "Many people think that innovation and tech is inherently expensive and is the primary reason for docs being 'slow adopters.' I say 'hooey'!"

Dr. Neu built his own practice website using a set of Apple tools, online at www.neucare.net. He uses a free, cloud-based electronic medical record (EMR), offers his patients email capabilities both through a secure site and—after they sign a disclaimer—through personal email and uses a secure version of Jotforms so that he can accept secure attachments such as blood sugar or blood pressure logs. His patients can request appointments online, and he will even do appointments via WebEx when the situation calls for something more than an email exchange but less than an office visit.

Because Dr. Neu uses a cloud-based EMR, he can look up a patient's record from anywhere and can easily handle prescription refills. He coaches his patients—especially those with chronic illnesses—to follow up with him, teaching them that he can care for them better if they have an ongoing relationship. Dr. Neu feels this approach is starting to have an impact on his patients with chronic conditions, who seem to feel more ownership in and more knowledge about their conditions.

Dr. Neu's practice is still young, but it is growing, and he is confident that his high tech *and* high touch brand of care will catch on.

Case Study 18

Total Access Physicians in Burlington, Kentucky

By Jan Oldenburg, FHIMSS (From an Interview with Gerry L.Tolbert, MD)

Dr. Gerry Tolbert practices medicine with his father, Jerry Tolbert, MD, at Total Access Physicians in Burlington, Kentucky. They are building a concierge practice that will be supported entirely by monthly fees.

In this practice, the doctors do all of their own lab work. Each has a second line on his cell phone dedicated to handling patient calls. They also provide patients with the use of a secure patient portal associated with their cloud-based practice management system.

Their patient portal allows them to provide patients with lab results, as well as other secure information about their conditions, enables direct patient scheduling, and provides secure email and text messages. It also has a smart phone application so that patients can get access to their information while mobile. Their iPads are loaded with patient education applications with anatomical drawings and physiology videos, and they have a library of reference materials for patients.

Dr. Tolbert talked about managing a patient with an abscess while at a conference. He used his PC to review the patient's email, access the electronic medical record (EMR), review the patient chart, and send a prescription—all while in the hallway outside the lecture room! Without the convenience of a remote access to the patient's chart, all of that would have required the patient to come in for an office visit—after the patient waited until he was back from the conference.

Dr. Tolbert is looking at how to measure and assess the outcomes of this new model. He is tracking how many times people call back for exacerbations of their conditions, as well as reviewing how people are doing with medications and compliance. They are seeing direct benefits in the form of increased patient trust and stronger patient relationships. He and his father see satisfaction almost directly mirrored in the patient recruitment effort, with 32% of clients coming in as a direct result of patient referrals. Their drop-out rate has been very low, and since they do not lock patients in for a term of service, it's a good proxy for patient satisfaction.

Case Study 19

Patient Engagement and Direct Practice

By Jan Oldenburg, FHIMSS (From an Interview with Josh J. Umbehr, MD)

At Atlas MD in Wichita, Kansas, Dr. Josh Umbehr believes that it makes good business sense to build a system with "no nos." His patients pay a monthly fee based on their age. Every procedure he can complete in the office is complimentary, and he offers everything from home visits to virtual visits, including text messages, direct Twitter responses, Facebook direct messages, cellphone conversations, Skype, and email—in short, he'll respond using any communication path that works for his clients. Patients purchase most medicines from Atlas MD at wholesale prices—and the savings are sometimes enough to cover their total monthly fee. Dr. Umbehr is trying to align all of the incentives so that doing the right thing is also the most cost-effective thing.

Atlas MD has two doctors and a nurse. They answer their own phones and focus on customer service and responsiveness. As one of his patients said, "I like best the fact that there are only two moving parts in this system: us and them."

One of the things that sets Dr. Umbehr apart is his flexible modes of communication. He's savvy about social media and uses Twitter (@atlasmd) and Facebook extensively for both marketing and communication. Dr. Umbehr observes that doctors have been slow to adopt social media and notes that the American Medical Association (AMA) was against the use of telephones when they first were introduced. He and his partner ask patients to sign waivers up front regarding the communication channels they use, signifying that they are aware that the channels are not totally secure. The following list shows, in descending order of frequency, the pathways Dr. Umbehr uses to communicate with patients:

- Email
- Text messages
- Twitter and Facebook
- Telephone
- Office visits
- Home visits (a distant last)

He told the story of being up late one night, contemplating a slice of cold pizza when his cell phone buzzed. One of his patients was traveling in China and sent him a direct Twitter message with a question. Josh responded within minutes, and his patient was able to happily continue his journey.

As for that practice of saying that any service they can perform in the office is complimentary, Dr. Umbehr noted, "When I am doing a laser hair removal on one of my female patients, and she's lying there relaxed, doing something for herself—you'd be surprised how often she'll ask a health question that just wouldn't have come up otherwise—and I get the opportunity to catch something early!"

While it is taking time to build their practice, after two years, they are on their way to profitability, with a set of patients who love the personal touch the practice is able to provide!

Case Study 20

Good Days Ahead, Online Cognitive Behavioral Therapy

By Eve Phillips, M.Eng, MBA

Good Days Ahead (GDA) from Empower Interactive, a digital behavioral health company, is an evidence-based, online psycho-educational program that teaches individuals the techniques of cognitive-behavior therapy (CBT) to address depression, anxiety, and stress in an engaging, interactive format. This program has been extensively tested as an adjunct to traditional clinical care, enabling the therapist to focus on high-value interactions instead of education and thus relieve stress on scarce clinical resources. It provides the added benefit of enabling patients to engage in their care directly through the GDA experience. The program also offers clinician and administrative interfaces to monitor and track individual progress. Most importantly, GDA has been designed with technical safeguards to protect end-user privacy and ensure Health Insurance Portability and Accountability Act (HIPAA) compliance.

The program's content was originally authored by Drs. Jesse Wright, Professor of Psychiatry at the University of Louisville, and Aaron Beck, University Professor Emeritus of Psychiatry at University of Pennsylvania, who was generally considered the father of cognitive therapy. Today, the program offers up to nine lessons, each of which is each designed to take approximately 30 minutes. During each lesson, individuals are invited to apply CBT concepts to their own life. The concepts are illustrated through videos and through interactive exercises including a Thought Change Record and an Activity Schedule. GDA's acceptance and efficacy has been studied in a number of research papers, demonstrating depression symptom reduction equivalent to in-person therapy, with half the investment of direct therapist time.[1,2]

GDA has been designed as a psycho-educational tool to support clinical care or as self-help, but is not, in itself, therapy. Its medical disclaimer reinforces this.[3] Empower has just submitted a mobile version of the Good Days Ahead app to the Apple Store.

Adoption and Engagement of Good Days Ahead in a Healthcare Organization

A large integrated healthcare organization was contracted to use GDA in fall 2011 to provide 10,000 individual uses of GDA for its members and their families at multiple facilities. They are using GDA as a supplement to clinical care, helping extend scarce clinician resources and improve access to evidence-based techniques. The agreement included associated clinical and administrative access and training.

The organization has placed GDA in several facilities where it is used across the spectrum behavioral health settings. Each setting offers unique opportunities to test the adoption of online behavioral health tools. To help drive end-user engagement, Empower provided the clinician community with tools, training, and written collateral (including brochures and notepads) to help them engage their patients in using GDA. The program is designed to drive both patient and clinician interactivity and adoption, and the trust relationship between clinicians and patients can be leveraged to encourage patient engagement in the program.

Outpatient: Clinicians in outpatient settings have used GDA as an adjunct to their traditional care for patients who show symptoms of stress, anxiety, or depression whom the clinician believes may benefit from an online tool. In particular, they have noted that those individuals with high anxiety and high levels of curiosity regarding their condition are especially likely to engage actively with GDA. Further research on integrating motivational interviewing techniques into the program, which work on facilitating and engaging intrinsic motivations, is underway.

Patient engagement comes in part through the individual's connection with the program's main character, "Lisa." In the program, patients learn through videos of Lisa, who puts the techniques in GDA to work to help her combat her depression. This process can take some time, however, and one clinician commented: "At first my most successful patient didn't really see the connection between the program and his experiences. But after Lesson 3, he became invested in finding out how the character overcame her depression and anxiety and then just flew through the lessons."

The connection with the main character, as well as the direct discussion about experiences common to those with anxiety or depression, can also help foster a feeling of hopefulness, as well as prevent the isolation so common for individuals with these issues. As one patient noted, "I don't feel alone anymore. I thought I was the only one who had these thoughts, but the fact that [my organization] made this huge effort to [offer] this program tells me that there must be a lot of people who are like me. It tells me I can do something about it, and that I'm normal."

Inpatient and Intensive Outpatient: Inpatient and intensive outpatient settings have implemented GDA into clinical processes with netbooks or kiosks that can access the program. For example, all patients receiving behavioral health treatment in an intensive outpatient setting at one clinic location are automatically signed up for a GDA account and assigned a netbook or time at a kiosk to complete their exercises over a period of four weeks. For these patients, the program helps provide a baseline of CBT skills to help them manage their moods and improve their ability to heal.

Intensive outpatient clinicians are able to effectively monitor the progress of this population and quickly identify who is progressing and who is not. This process ensures that clinical resources can be allocated as effectively and efficiently as possible to allow clinicians to provide additional direct services to those who may need more assistance. Additional clinicians can also be securely assigned to individual accounts as necessary when new clinicians join or additional clinical insight is requested.

Group Therapy Programs: In group therapy settings where clinicians use CBT principles as part of their methodology, the GDA program has been used to ensure that new members have an opportunity to learn the principles at their own speed, independent from the flow of the group. In groups where individuals all begin at the same time, the group sessions can be aligned with the lessons in the program. These lessons have also provided potential topics for the group discussion. Using GDA, patients' engagement in their therapeutic process can stretch beyond the group therapy session into life outside of the group room.

Future Use: Primary Care: Primary care is a key setting in the effort to expand access to behavioral health services. Unfortunately, behavioral health professionals frequently note that many of the patients who express behavioral health issues in primary care often do not go on to receive help from behavioral health specialists. In these cases, GDA can be used by primary care clinicians to ensure that those patients can receive a baseline of CBT skills training. This tool thus offers patients an opportunity to deal with behavioral healthcare issues without the commitment required to see a specialist. In cases in which there are wait lists for behavioral health specialists, GDA can also serve as an introduction to CBT techniques that can be continued in more depth once patients are able to meet with a clinician.

Benefits of Program Use

Provide structured framework for efficient, effective treatment: GDA provides a framework for teaching CBT principles and gives structure for patient treatment. Because the program enables a clinician to track patients' progress and review completed exercises prior to their scheduled appointment, the clinician is armed with additional knowledge to "jump start" talk therapy sessions with his or her patients. This ensures the clinician's time with patients is used as efficiently as possible.

Increase access to evidence-based care: Depression and anxiety symptoms are common in the general population and can be present both on their own and in conjunction with a variety of other behavioral and physical disorders, including post-traumatic stress disorder (PTSD). CBT is an evidence-based therapy approach to help individuals manage those conditions. Any individual with these symptoms is a potential candidate for using GDA. Ultimately, it is up to the clinician to determine whether a patient is suited to the program.

Reduce burden of demand for behavioral health services: Clinicians (as supported by published clinical trial data) have suggested that GDA be deployed to patients when they are first referred for behavioral health services and triage between referrals and appointments to reduce the burden of wait times for behavioral health sessions.

Support continuity of care: The GDA program provides a consistent, convenient, and quantified experience for patients that can also help assist in ensuring continuity of care. If patients change clinicians or relocate, they can continue usage of the program and access to their own data anywhere there is Internet access, regardless of whether they continue to receive treatment. Use of this program enables continuity of care across the healthcare enterprise as a whole by ensuring data fidelity across multiple stakeholder groups (individuals, clinicians, families) and by removing the responsibility of transporting paperwork from individuals and families.

Drive patient engagement: Ultimately, this program can help patients manage their own learning process that is key to the cognitive-behavior therapy methodology for key behavioral health issues. Instead of depending on sessions with a clinician, patients can take their health into their own hands in this interactive e-learning environment. The program empowers as it teaches, for its intent is to provide patients with evidence-based tools to successfully manage their moods and behaviors.

REFERENCES

1. Wright JH, Wright AS, Salmon P, et al. Development and initial testing of a multimedia program for computer-assisted cognitive therapy. *Am J Psychother.* 2002; 56(1):76-86.

2. Wright JH, Wright AS, Albano AM, et al. Computer-assisted cognitive therapy for depression: maintaining efficacy while reducing therapist time. *Am J Psychiatry.* 2005 Jun; 162(6):1158-64.

3. Medical Disclaimer: The *Good Days Ahead* website, including all information, contents, materials, services and links contained therein, is to be used for informational purposes only. It is not intended to be a substitute for professional medical advice, diagnosis, or treatment. Individuals should use their own judgment and common sense when using the website, and they should always seek the advice of their physician or other qualified health provider with any questions they may have regarding a physical or mental condition.

Case Study 21

Patient Portals Offer Opportunities for Patient-Provider Interaction at Children's Hospital of Philadelphia

By Deborah C. Wells, MS, CPHIMS

The Children's Hospital of Philadelphia (CHOP) is a 430-bed pediatric hospital. Last year, it had more than 1 million outpatient and inpatient visits. Children's Hospital has a growing network of facilities and resources created to provide exceptional services to all children for every possible need.

Pediatricians play an important role in a patient's transition from child to adult. This process is often viewed with conflicting emotions by the parent. The first time that a boy does not want his mother in the exam room with him can be a little traumatic for Mom. CHOP clinicians help families work through this challenging time in a child's life.

When CHOP implemented a patient portal, the confidentiality rights of adolescents were an important consideration. When a child enters adolescence, state and federal laws afford competent minors confidentiality rights that allow them to receive certain services without their parents' consent. The adolescent, in such a case, also controls access to health information related to that service. Covered services often include pregnancy and reproductive issues, substance abuse, HIV (human immunodeficiency virus), and mental health.

Laws governing adolescent privacy are complex and nuanced. In addition to privacy for special conditions, Pennsylvania law allows "emancipated minors" to control access to all of their health information. The law stipulates the conditions under which a child may seek emancipated status. These conditions include marriage, pregnancy, educational status, and others.

Since the electronic health record (EHR) does not have functionality to analyze a patient's status regarding complex legal criteria, CHOP faced the task of defining portal policies that would cover all of the laws' provisions.

One challenge was to understand where in the EHR sensitive information could be found. For some types of information, patient-controlled data may be mixed with non-patient-controlled data. If parents have access to the medication list, for example, a confidential service received by a minor could, inadvertently, be revealed. Other types of potentially confidential service information can include lab results, billing information, and visit types.

- Although all lab results are documented in the same place in the EHR, CHOP was able to flag some labs for release to the portal, but restrict access to others that were sensitive. The task was relatively straightforward, albeit labor intensive, since each lab type had to be flagged individually.

- Medications are also documented in a single list. In this case, however, it was not possible or workable to review and flag all relevant medications. Medications are particularly difficult since some of them can be used to treat more than one problem. A medication cannot be flagged when it is confidential if prescribed for one condition, but not when used for other conditions.

- Visit reminders include the type of service and the name of the provider with whom the visit is scheduled. This information could inadvertently reveal the condition for which the patient is being treated. Unfortunately, functionality for sending visit reminders via the portal is an "all or nothing" setting in our system.

Another problem was determining how to restrict parental access to controlled information while allowing them access to the rest of their child's record.

In the end, CHOP decided not to try to implement complex criteria tailored to each type of information, particularly since such criteria might or might not protect all of the privileged information. Instead, the approach decided upon was to select an age at which the adolescent, not the parent, would control access to the relevant parts of the health record. Both legal analysis and CHOP's own experience and policies indicated that the age of 13 would be appropriate.

Just after midnight on a child's 13th birthday, his or her parents' access to information that could directly or indirectly reveal confidential information is automatically restricted. In explaining this to the family, the physician has an opportunity to discuss adolescents' increasing responsibility for managing their own health.

Protecting the privacy of our adolescent patients is a challenging, but important feature of our patient portal. Technology, we know, can have unintended consequences. Who knew that a patient portal would be a tool for focusing parents, patients, and providers on what "becoming an adult" means and defining the role of information privacy in that transition.

Lessons Learned

Because the portal is patient-facing, its implementation requires input from a more highly-varied, multi-disciplined team than do most other projects. Clinical, ancillary, legal, and administrative representatives from across the institution, as well as patient and family advocates, all have important contributions to make.

Providing electronic information to users outside of the institution presents different issues than information systems that are used exclusively within the organization. Determining the unique needs of the external audience and how to meet them is a challenging but essential task.

Case Study 22

Veterans Health Administration, Department of Veterans Affairs– Patient Engagement and the Blue Button

By Susan Woods, MD, MPH

The 2001 Institute of Medicine (IOM) seminal report, *Crossing the Quality Chasm*, recommended fostering healthcare system innovation using consumer health information technology to transform the delivery of care. In 1999, prior to the IOM report and proving prescient for its time, the U.S. Department of Veterans Affairs (VA) began developing and piloting a personal health record (PHR). My HealtheVet was one of the first PHRs in the country, a testament to VA's forward-thinking approach to engaging Veteran patients. The initial PHR was piloted with 7,500 patients and provided electronic access to clinical notes, hospital discharge summaries, lists of allergies and medication(s), and laboratory and imaging results. Surveys and interviews with patient and family users revealed the response to full sharing of electronic health records (EHR) was overwhelmingly positive. Respondents noted they had a greater understanding of their health conditions, improved communication with providers, and enhanced involvement in their care.[1, 2]

In 2003, the My HealtheVet PHR transitioned to a national platform and now serves as the cornerstone of VA's electronic services that engage patients (available online at: http://www.myhealth.va.gov). The primary goal of My HealtheVet is to support Veterans as empowered healthcare consumers through improved quality, access and value of healthcare services, as well as increased satisfaction.[3] Over time, features have been added to My HealtheVet, affording users the ability to securely email their VA health care team, refill medication(s), view wellness reminders, track vitals and metrics, and have access to trusted health information; journaling; and calendar functions. Tailored reminders are generated from VA's EHR for immunizations, lipid and diabetes testing, elevated blood pressure, and body mass index, and screenings for colon; breast; and cervical cancer. EHR data currently available to authenticated users includes the ability to access and view medication(s), hematology and chemistry test results, allergies, immunizations, as well as appointments.

The ability to save and share personal health data was also identified as an important need for patients, based on survey responses from My HealtheVet website visitors.[3,4,5] In response, VA Blue Button functionality was launched in August 2010.[6] Using VA Blue Button, Veterans can download My HealtheVet personal health information as a PDF, text file, or a .bluebutton file. The data file can then be printed or saved to a personal computer or portable device.

As of September 2012, almost two million Veterans had registered for My HealtheVet, with 896,369 VA patients authenticated to use all functions. Over 33 million medication refills have been requested online, and 439,119 Veterans have opted in to use Secure Messaging, an online patient-provider communication functionality. More than 610,000 patients have used the VA Blue Button download for more than 2.4 million data files (58% as PDF, 25% as .bluebutton, and 17% as text files). My HealtheVet users mirror the population of Veterans using VA health services overall; approximately 68% of users are 51 to 70 years of age, and 16% are 71 years and older. Among the 88,591 My HealtheVet visitors who were not first-time users, 91% reported using the website monthly or more frequently.[3] Prior to having access to Secure Messaging, 56% felt that My HealtheVet improved their ability to manage their health. With the advent of Secure Messaging across VA primary care during 2011, patient satisfaction with the PHR and its role in enhancing access and communication is believed to be significantly higher.

VA continues to expand access to personal health information and develop digital services for Veterans and their caregivers. Resources have also been made available to support the technology's reach across the VA population. Each medical center boasts a full-time My HealtheVet Coordinator serving as a PHR champion and an invaluable liaison between patients, caregivers, and clinical and administrative staff. As the online tools grow more robust, the coordinators become even more integral to VA's vision of providing high-quality, patient-centered care. Development continues across Internet, mobile, and point-of-service kiosk platforms. The full EHR information that was available for the PHR Pilot, including clinical notes and hospital discharge summaries, will soon be available through the My HealtheVet website and VA mobile Blue Button. Feedback from patients, families, and clinicians, along with rigorous research and evaluation, are continuously incorporated into VA's eHealth initiative to bring services of the highest value to Veterans across the country everywhere.

Veterans talk about the My HealtheVet Pilot —with full access to notes

"You can look back at your records and see, you know, a few years ago, did I have the same problem or I think I did this or I think I did that?"

"I can go in and ask more intelligent questions and we don't have to spend as much time with them explaining everything to me. And then with my stress level up at the doctor's office, I don't hear half of it and then we may have to do it again and again. And so, it helps us to have better communication."

"I think it added years on to my life…because I know more."

"You know, the more information that I had, the better I felt, whether it was good or bad."

"Personally, it helped me assume the role of taking care of my own health which my wife, a nurse, said, 'I'm not taking care of you anymore. You've got to take care of you.' So, all of a sudden, I had access to the information so I could do that."

VA Blue Button

 The VA Blue Button is available for use on My HealtheVet and enables Veterans to download, view, store and print information from their personal health record.

REFERENCES

1. Nazi KM, Hogan TP, McInnes DK, et al. Evaluating patient access to electronic health records: results from a survey of veterans. *In press.*

2. Woods SS, Schwartz E, Tuepker A, et al. Patient experiences with full electronic access to health records and clinical notes through the My HealtheVet Personal Health Record Pilot: a qualitative study. *In press.*

3. Nazi KM. Veterans' voices: use of the American Customer Satisfaction Index (ACSI) Survey to identify My HealtheVet personal health record users' characteristics, needs and preferences. *J Am Med Inform Assoc.* 2010;17:203-211.

4. Zulman DM, Nazi KM, Turvey CL, et al. Patient Interest in sharing personal health record information: a web-based survey. *Ann Intern Med.* 2011; 155:805-810.

5. Carolyn L. Turvey CL, Zulman DM, et al. Transfer of information from personal health records: a survey of veterans using My HealtheVet. *Telemedicine eHealth.* 2012; 18(2):109-114.

6. *Accessing your health information with the Blue Button.* U.S. Department of Health and Human Services. http://www.healthit.gov/bluebutton. Accessed October 14, 2012.

Case Study 23

Prescribing Apps for Weight Loss

By Brad Tritle, CIPP, Bianca K. Chung, MPH, and Karen Colorafi, RN, MBA, BScN, CPEHR, CPHIT (From an Interview with Michael H. Zaroukian, MD, PhD, FACP, FHIMSS)

Dr. Michael Zaroukian, an internist practicing in East Lansing, MI, has found an innovative way to use mobile health with a handful of his patients. One day at the end of a busy clinic, Dr. Zaroukian found himself engaged in a conversation with a patient who was frustrated by his inability to lose weight. Dr. Zaroukian explained, "He had a desire to lose weight but had difficulty connecting his actions with his weight loss efforts. Since we're near the Motor City, I mentioned it is like driving; you need to always keep your eyes on the road!" Dr. Zaroukian noticed that the patient had an iPhone (as did the doctor) and so the two of them spent time looking through the app store for something that would help. Together, they found *Lose It!* and were impressed by its high ratings and the sheer volume of reviews. It took less than three minutes to examine the free app and download it to their phones.

The functionality of *Lose It!* is really exciting. Patients can enter basic profile information, set weight loss goals, chart foods eaten (saving commonly consumed foods in a favorite list for repeat entry), scan bar codes of packaged foods, track exercise, calculate daily caloric requirements for their specific weight goal and even securely share their progress with friends. Dr. Zaroukian boldly promised to also try out the app and let his patient track his progress—a commitment he has since made to other patients. Dr. Zaroukian's only rules about weight loss were to use the app, stay within the "calorie budget" for the day set by their weight loss goal, and avoid eating only those foods that they could not resist binging on. The patient has lost over 20 pounds so far, and eleven other patients are connected to the doctor when he logs onto the app each day.

The app is the "tipping point," Dr. Zaroukian explained, the thing that gives patients a new sense of power, control, and allows them to connect their behaviors and activities—eating and exercise—with their weight loss progress. Dr. Zaroukian strikes a careful balance between using the social-media functions of the app to cheer patients on while using secure messaging functionality at the office for answering clinical questions or revealing protected health information. Although this is a small sample, it appears that logging and tracking information regularly is correlated with weight loss for his group of patients. The social support, nurturing, and accountability seems to help too, in the same way that Weight Watchers helps. Dr. Zaroukian prescribes the *Lose It!* app to patients at the clinic on a regular basis now, in addition to others such as *dood Stretch* and *Fooducate*. He estimates that this approach isn't any more time-consuming than handing out educational material and explaining it. Instead, he said, "These apps engage the patients and are more effective than the handouts I used to distribute. This is a part of the patient's care, part of their therapy, so I think about it in terms of doing what's best for the patient."

Dr. Zaroukian hopes that medical apps get better and better, potentially barcoding prescriptions so that it's easier to track medication adherence and watch out for drug-drug interactions in the same way that *Lose It!* allows you to track the relationship between food intake, exercise, and weight loss. When his patients ask how long they should continue to use *Lose It*, he replies, "How long do you have to look at the road when you're driving? The answer is all the time unless they want to drive off the road or crash. I remind them that successful long-term weight management is a journey, not a destination."

Appendix A

Personal Health IT Tools and Categories

Category	Patient Engagement Capability
Communication	Communicate securely with Provider or Care Team
	Complete a virtual visit (extended email or video)
	Receive and schedule alerts and reminders
Convenient Self-Service	Find a physician or facility
	Make, change, or cancel an appointment
	Check-in and register for a visit
	Manage prescription refills
Personal Health Information	Review ambulatory clinical Information (MU)
	Review inpatient clinical information (MU)
	Download or transmit your health record (MU)
	Enter or track personal data
	Upload data from devices
	Review medication information
Financial	View clinical bills
	Pay clinical bills
	Estimate the cost of services
	Manage HSA/HRA/FSA* accounts
	Reconcile bills and claims
Education and Support	Manage chronic illness
	Make decisions about care or treatment
	Educate yourself about health issues
	Receive peer and community support

Category	Patient Engagement Capability
General Capabilities	Manage identity and access to your account
	Obtain administrative support
	Manage your profile and preferences
	Manage insurance and coverage
	Analyze your health and health data

* HSA is healthcare savings account; HRA is healthcare reimbursement account; FSA is flexible spending account.

Appendix B

Meaningful Use Measures

Stage 1 Meaningful Use Measures—
Engage Patients and Their Families

MU Category	EP	EH	Objective	Measure
Core: Electronic copy of Health Information	X	X	**Eligible Professional and Hospital:** Provide patients with an electronic copy of their health information (including diagnostic test results, problem list, medication lists, medication allergies), upon request.	**Eligible Professional and Hospital:** More than 50% of all patients of the EP or the inpatient or emergency departments of the eligible hospital (EH) or critical access hospital (CAH) (POS 21 or 23) who request an electronic copy of their health information are provided it within 3 business days.
Core: Electronic copy of Discharge Instructions		X	**Eligible Hospital:** Provide patients with an electronic copy of their discharge instructions at time of discharge, upon request.	**Eligible Hospital:** More than 50% of all patients who are discharged from an eligible hospital or CAH's inpatient department or emergency department and who request an electronic copy of their discharge instructions are provided it.
Core: Clinical Summaries	X		**Eligible Professional:** Provide clinical summaries for patients for each office visit.	**Eligible Professional:** Clinical summaries provided to patients for more than 50% of all office visits within 3 business days (paper or electronic).
Menu: Electronic access to health data	X		Provide patients with timely electronic access to their health information (including lab results, problem list, medication lists, medication allergies) within four business days of the information being available to the eligible professional.	More than 10% of all unique patients seen by the eligible professional are provided timely (available to the patient within four business days of being updated in the certified EHR technology) electronic access to their health information subject to the eligible professional's discretion to withhold certain information.
Menu: Patient-specific education	X	X	Use certified EHR technology to identify patient-specific education resources, and provide those resources to the patient if appropriate.	More than 10% of all unique patients seen by the eligible professional or admitted to the eligible hospital's inpatient or emergency department during the EHR reporting period are provided patient-specific education resources.

Stage 2 Meaningful Use Measures— Engage Patients and Their Families

MU Category	EP	EH	Objective	Measure
Core: Reminders	X		**Eligible Professional:** Use clinically relevant information to identify patients who should receive reminders for preventive/follow-up care, and send these patients a reminder, per patient preference.	**Eligible Professional:** More than 10% of all unique patients who have had two or more office visits with the EP within the 24 months before the beginning of the EHR reporting period were sent a reminder, per patient preference when available.
Core: Patients can view online, download, and transmit their health records.	X		**Eligible Professional:** Provide patients the ability to view online, download, and transmit their health information within 4 business days of the information being available to the EP.	**Eligible Professional:** 1. More than 50% of all unique patients seen by the EP during the EHR reporting period are provided timely (within 4 business days after the information is available to the EP) online access to their health information subject to the EP's discretion to withhold certain information. 2. More than 5% of all unique patients seen by the EP during the EHR reporting period (or their authorized representatives) view, download, or transmit to a third party their health information. Specific information that an eligible professional must make available to the patient includes • Patient name • Provider's name and office contact information • Current and past problem list • Procedures • Laboratory test results • Current medication list and medication history • Current medication allergy list and medication allergy history • Vital signs (height, weight, blood pressure, body mass index [(BMI)], growth charts) • Smoking status • Demographic information (preferred language, sex, race, ethnicity, date of birth) • Care plan field(s), including goals and instructions • Any known care team members including the primary care provider (PCP) of record

MU Category	EP	EH	Objective	Measure
Core: Patients can view online, download, and transmit their health records.		X	**Eligible Hospital:** Provide patients the ability to view online, download, and transmit information within 36 hours after discharge from the hospital.	**Eligible Hospital:** 1. More than 50% of all patients who are discharged from the inpatient or emergency department (POS 21 or 23) of an eligible hospital or CAH have their information available online within 36 hours of discharge 2. More than 5% of all patients (or their authorized representatives) who are discharged from the inpatient or emergency department (POS 21 or 23) of an eligible hospital or CAH view, download or transmit to a third party their information during the reporting period. The information that an eligible hospital must provide to satisfy the objective and measure: • Patient name • Admit and discharge date and location • Reason for hospitalization • Care team including the attending of record as well as other providers of care • Procedures performed during admission • Current and past problem list • Current medication list and medication history • Current medication allergy list and medication allergy history • Vital signs at discharge • Laboratory test results (available at time of discharge) • Summary of care record for transitions of care or referrals to another provider • Care plan field(s), including goals and instructions • Discharge instructions for patient • Demographics maintained by hospital (sex, race, ethnicity, date of birth, preferred language) • Smoking status
Core: Clinical Summaries	X		**Eligible Professional:** Provide clinical summaries for patients for each office visit.	**Eligible Professional:** Clinical summaries provided to patients or patient-authorized representatives within 1 business day for more than 50% of office visits.
Core: Secure Messaging	X		**Eligible Professional:** Use secure electronic messaging to communicate with patients on relevant health information.	**Eligible Professional:** A secure message was sent using the electronic messaging function of CEHRT by more than 5% of unique patients (or their authorized representatives) seen by the EP during the EHR reporting period.
Core: Patient-specific education	X	X	**Eligible Professional and Eligible Hospital:** Use clinically relevant information from Certified EHR Technology to identify patient-specific education resources, and provide those resources to the patient if appropriate.	**Eligible Professional:** Patient-specific education resources identified by CEHRT are provided to patients for more than 10% of all unique patients with office visits seen by the EP during the EHR reporting period. **Eligible Hospital:** More than 10% of all unique patients admitted to the eligible hospital's or CAH's inpatient or emergency departments are provided patient-specific education resources identified by Certified EHR Technology.

Stage 3 Meaningful Use Recommendations—
Engage Patients and Their Families

ID	Stage 2 Final Rule	Stage 3 Recommendations	Future Possibilities/ Comments
SGRP 116 Reminders	**EP Objective:** Use clinically relevant information to identify patients who should receive reminders for preventive/ follow-up care, and send these patients the reminder per patient preference. **Measure:** More than 10% of all unique patients who have had two or more office visits with the EP within the 24 months before the beginning of the EHR reporting period were sent a reminder, per patient preference when available.	**EP Objective:** Use clinically relevant information to identify patients who should receive reminders for preventive/follow-up care. **EP Measure:** More than 20% of all unique patients who have had an office visit with the EP within the 24 months prior to the beginning of the EHR reporting period were sent a reminder, per patient preference. **Exclusion:** Specialists may be excluded for prevention reminders (could be more condition specific).	
SGRP 120 Electronic notes	**EP/EH MENU Objective:** Record electronic notes in patient records. **EP MENU Measure:** Enter at least one electronic progress note created, edited and signed by an eligible professional for more than 30% of unique patient office visits. **EP MENU Measure:** Enter at least one electronic progress note created, edited, and signed by an authorized provider of the eligible hospital's or CAH's inpatient or emergency department (POS 21 or 23) for more than 30% of unique patients admitted to the eligible hospital or CAH's inpatient or emergency department during the EHR reporting period.	Record electronic notes in patient records for more than 30% of office visits within four calendar days.	

ID	Stage 2 Final Rule	Stage 3 Recommendations	Future Possibilities/ Comments
SGRP 204A	**EP Objective:** Provide patients the ability to view online, download, and transmit (VDT) their health information within 4 business days of the information being available to the EP. **EP Measure:** 1. More than 50% of all unique patients seen by the EP during the EHR reporting period are provided timely (within 4 business days after the information is available to the EP) online access to their health information subject to the EP's discretion to withhold certain information. 2. More than 5% of all unique patients seen by the EP during the EHR reporting period (or their authorized representatives) view, download, or transmit to a third party their health information. **EH Objective:** Provide patients the ability to view online, download, and transmit information about a hospital admission 1. More than 50% of all patients who are discharged from the inpatient or emergency department (POS 21 or 23) of an eligible hospital or CAH have their information available online within 36 hours of discharge 2. More than 5% of all patients (or their authorized representatives) who are discharged from the inpatient or emergency department (POS 21 or 23) of an eligible hospital or CAH view, download or transmit to a third party their information during the reporting period.	EPs should make info available within 24 hours if generated during course of visit. For labs or other types of information not generated within course of visit, it is made available to patients within four business days of information becoming available to EPs. **MENU item:** Automated Transmit*: (builds on Automated Blue Button Initiative [(ABBI)]): Provide 50% of patients the ability to designate to whom and when (i.e. pre-set automated & on-demand) a summary of care document is sent to patient-designated recipient** *Subject to the same conditions as view, download, transmit. **Before issuing final recommendations in May 2013, HITPC will also review the result of Automated Blue Button pilots, in addition to considering public comments received.	Create capability for providers to review patient-transmitted information and accept updates to the EHR. Explore possibility of including • Actual images • Radiation dosing **MENU item:** Consider making a menu item to make electronic notes available to patients aka the OpenNotes project.

ID	Stage 2 Final Rule	Stage 3 Recommendations	Future Possibilities/ Comments
SGRP 204B	New	**MENU:** Provide 10% of patients with the ability to submit patient-generated health information to improve performance on high-priority health conditions, and/or to improve patient engagement in care (e.g. patient experience, pre-visit information, patient created health goals, shared decision making, advance directives, etc.). This could be accomplished through semi-structured questionnaires selected by the EP or EH.	**Explore** readiness to include medical device data from the home. How can the HITECH incentive program support allowing doctors and patients to mutually agree on patient-generated data flows that meet their needs, and should the functionality to collect those data be part of EHR certification?
SGRP 204D	New	**Objective:** Provide patients with the ability to request an amendment to their record online (e.g., offer corrections, additions, or updates to the record) online.	
SGRP 205	**EP Objective:** Provide clinical summaries for patients for each office visit. **EP Measure:** Clinical summaries provided to patients or patient-authorized representatives within 1 business day for more than 50% of office visits.	The clinical summary should be pertinent to the office visit, not just an abstract from the medical record.	What specific information should be included in the after visit summary to facilitate the goal of patients having concise and clear access to information about their most recent health and care and understand what they can do next, as well as when to call the doctor if certain symptoms/events arise?
SGRP 206	**EP/EH Objective:** Use Certified EHR Technology to identify patient-specific education resources, and provide those resources to the patient. **EP CORE Measure:** Patient-specific education resources identified by Certified EHR Technology (CEHRT) are provided to patients for more than 10% of all unique patients with office visits seen by the EP during the EHR reporting period. **EH CORE Measure:** More than 10% of all unique patients admitted to the eligible hospital's or CAH's inpatient or emergency departments (POS 21 or 23) are provided patient-specific education resources identified by Certified EHR Technology.	**Additional language support:** For the top 5 non-English languages spoken nationally, provide 80% of patient-specific education materials in at least one of those languages based on EP's or EH's local population, where publically available.	

ID	Stage 2 Final Rule	Stage 3 Recommendations	Future Possibilities/ Comments
SGRP 207	**EP Objective:** Use secure electronic messaging to communicate with patients on relevant health information. **EP Measure:** A secure message was sent using the electronic messaging function of Certified EHR Technology by more than 5% of unique patients (or their authorized representatives) seen by the EP during the EHR reporting period.	**Measure:** More than 10% of patients use secure electronic messaging to communicate with EPs.	Create capacity for electronic episodes of care (telemetry devices, etc.) and to do e-referrals and e-consults.
SGRP 208	**Not included separately (in reminder objective)**	**EP and EH Measure:** Record communication preferences for 20% of patients, based on how (e.g., the medium) patients would like to receive information for certain purposes (including appointment reminders, reminders for follow up and preventive care, referrals, after visit summaries and test results).	
SGRP 209	New	**Certification Criteria:** Capability for EHR to query research enrollment systems to identify available clinical trials. No use requirements until future stages.	The goal of this objective is to facilitate identification of patients who might be eligible for a clinical trial, if they are interested.
SGRP 302	**EP/EH CORE Objective:** The EP/EH who receives a patient from another setting of care or provider of care or believes an encounter is relevant should perform medication reconciliation. **EP/EH CORE Measure:** The EP, eligible hospital, or CAH performs medication reconciliation for more than 50% of transitions of care in which the patient is transitioned into the care of the EP or admitted to the eligible hospital's or CAH's inpatient or emergency department (POS 21 or 23).	**EP / EH / CAH Objective:** The EP, eligible hospital or CAH who receives a patient from another setting of care or provider of care or believes an encounter is relevant should perform reconciliation for: - medications - medication allergies - problems **EP / EH / CAH Measure:** The EP, EH, or CAH performs reconciliation for medications for more than 50% of transitions of care, and performs reconciliation for medication allergies, and problems for more than 10% of transitions of care in which the patient is transitioned into the care of the EP or admitted to the eligible hospital's or CAH's inpatient or emergency department (POS 21 or 23). **Certification Criteria:** Standards work needs to be done to adapt and further develop existing standards to define the nature of reactions for allergies (i.e., severity).	Explore the feasibility of adding additional fields for reconciliation such as social history. Also explore potential for an electronic shared care planning and collaboration tool that crosses care settings and providers, allows for and encourages team-based care, and includes the patient and their non-professional caregivers.

Index

A

Accountable Care Organizations (ACO), 6, 9–17, 27, 66, 139
    and chronic conditions, 10
    features of, 9–10
    and patient satisfaction, 15
    and personal health IT, 10–13
Admissions, 81
    pre-registration process, 81
Affordable Care Act (ACA), 43, 144
Agency for Healthcare Research and Quality (AHRQ), 11, 49
American Health Information Management Association (AHIMA), 150
American Recovery and Reinvestment Act (ARRA), 22, 43, 143
Appointment scheduling, online, 75–77
    direct scheduling, 76
    reminder apps, 92
Association of Cancer Online Resources, 103
Automated Blue Button Initiative (ABBI), 145

B

Bedsider case study, 130–131
Blogs, 102–104, 109
Blue Button, 34, 52, 138, 145
    case study, 222–224
    security and privacy issues, 125–126

I

K

L

M